STARS AND STAR SYSTEMS

ASTROPHYSICS AND SPACE SCIENCE LIBRARY

A SERIES OF BOOKS ON THE RECENT DEVELOPMENTS
OF SPACE SCIENCE AND OF GENERAL GEOPHYSICS AND ASTROPHYSICS
PUBLISHED IN CONNECTION WITH THE JOURNAL
SPACE SCIENCE REVIEWS

VOLUME 75
PROCEEDINGS

STARS AND STAR SYSTEMS

PROCEEDINGS OF THE FOURTH EUROPEAN REGIONAL MEETING IN ASTRONOMY
HELD IN UPPSALA, SWEDEN, 7 - 12 AUGUST, 1978

Edited by

BENGT E. WESTERLUND

Astronomical Observatory, University of Uppsala, Sweden

Jointly sponsored by

THE INTERNATIONAL ASTRONOMICAL UNION
AND
THE ASTRONOMY AND ASTROPHYSICS DIVISION OF
THE EUROPEAN PHYSICAL SOCIETY

D. REIDEL PUBLISHING COMPANY

DORDRECHT : HOLLAND / BOSTON : U.S.A.
LONDON : ENGLAND

Library of Congress Cataloging in Publication Data

CIP

European Regional Meeting in Astronomy, 4th, Uppsala, 1978.
Stars and star systems.

(Astrophysics and space science library ; v. 75)
Bibliography: p.
Includes index.
1. Stars–Congresses. 2. Galaxies–Congresses. I. Westerlund, Bengt E.
II. International Astronomical Union. III. European Physical Society.
Astronomy and Astrophysics Division. IV. Title. V. Series.
QB799.E96 1978 523.8 79-11467
ISBN 90-277-0983-1

Published by D. Reidel Publishing Company,
P. O. Box 17, Dordrecht, Holland.

Sold and distributed in the U.S.A., Canada and Mexico
by D. Reidel Publishing Company, Inc.
Lincoln Building, 160 Old Derby Street, Hingham,
Mass. 02043, U.S.A.

Printed in The Netherlands

TABLE OF CONTENTS

PREFACE

The 4th European Regional Meeting in Astronomy, entitled "Stars and Star Systems", was held in Uppsala, Sweden, on August 7 - 12, 1978. It was attended by 228 participants from 24 countries. Over 100 papers were presented; this required parallel sessions throughout the Meeting.

Financial support was given by the IAU, the Swedish Department of Education, the Swedish Natural Science Research Council, the Swedish Institute and the University of Uppsala.

The members of the Scientific and Local Organizing Committees are listed below, and I thank all of them for their contributions to the Meeting.

The Meeting was divided into six sections, according to the scientific topics: Galaxies (A) - including galactic structure and star formation; High-Energy Astrophysics (B); Stars (C); Interstellar Processes (D); Astronomical Instrumentation (E) and Education in Astronomy (F). In each section a number of Invited Papers were presented as well as a large number of contributed papers. In addition, each day a General Lecture was given by an Invited Speaker.

The present volume contains all the General Lectures and all the Invited Papers presented during the Meeting. In three cases, however, and for various reasons, only brief summaries have been available. Abstracts of the contributed papers may be found in Uppsala Astronomical Observatory, Report No. 12.

The proceedings were made from camera-ready typescripts, prepared in Uppsala. In the manuscripts, submitted by the speakers, only minor corrections have been made. More effort has been put into getting the Proceedings ready for publication at the earliest possible date than into attaining complete homogeneity and perfection typographically.

November 1978

Bengt E. Westerlund
Editor

SCIENTIFIC ORGANIZING COMMITTEE

J. Appenzeller	FRG
C. Cesarsky	France
M. Hack	Italy
L. Houziaux	Belgium
E. Kharadze	USSR
F. Pacini	ESO
J.-C. Pecker	France
V.-C. Reddish	Great Britain
B. Strömgren	Denmark
J. Trümper	FRG
H. van der Laan	Holland
P.A. Wayman	Ireland
B.E. Westerlund	Sweden (Chairman)

LOCAL ORGANIZING COMMITTEE

B.E. Westerlund, Chairman
N.Å.S. Bergwall
A.B.G. Ekman
Inga Henriksson
L. Lundin

LIST OF PARTICIPANTS

A. P. Abbasov, Shemaka Astrophysical Observatory, Azerbaijan Academy of Sciences, Shemaka, U.S.S.R.
A. Achterberg, Scherpenzeelseweg 2, 3953 MB Maarsbergen, Holland.
A. W. Al-Sabti, Scientific Research Foundation, Astronomy Dept., Jadhria, Baghdad, Irak
H. Albers, Sterrewacht, Leiden, Holland
A. K. Alksnis, Radioastrophysical Observatory, Latvian Academy of Sciences, Riga, U.S.S.R.
L. Allamandola, Laboratory Astrophysica, Huygens Lab., Wassenaarseweg 78, Leiden 2405, Holland
W. J. Altenhoff, Max-Planck-Institut für Radioastronomy, Auf dem Hügel 69, D-5300 Bonn 1, F.R.G.
J. Andersen, Copenhagen University Observatory, Brorfelde, DK-4340 Töllöse, Denmark
I. Appenzeller, Landessternwarte, 6900 Heidelberg-Köningsstuhl, F.R.G.
A. Ardeberg, Inst. för Astronomi, Svanegatan 9, 222 24 Lund, Sweden
M. G. Arntzen, Inst. of Theor. Astrophys., Univ. of Oslo, P.O. Box 1029, Blindern, Oslo 3, Norway
B. Aschenbach, Max-Planck-Institut, 8046 Garching, F.R.G.
E. Asseo, Observatoire de Meudon, 921 90 Meudon, France
E. Athanassoula, Besançon Observatory, 41, Bis, Av. de l'Observatoire 25000 Besançon, France
D. J. Axon, Astron. Center, Univ. of Sussex, Palmer, Brighton, England
F. Baas, Laboratory Astrophysica, Huyghens Lab., Wassenaarseweg 78, Leiden 2405, Holland
B. A. Balazs, Dept. of Astronomy, L. Eötvös Univ., Kun Bela T. 2, H-1083 Budapest, Hungary
L. Balazs, Konkoly Observatory, Box 67, 1525 Budapest, Hungary
R. Barbon, Osservatorio Astrofisico, I-36012 Asiago, Italy
N. Barylko, Inst. of Phys. Univ. of Gdansk, Ul Wita Stwosza 57, 80-952 Gdansk, Poland
E. Basinska-Grzesik, Pol. Acad. of Sciences Cop. Astr. Center, Chopina 12-18, 87-100 Torun, Poland
P. J. Bedijn, Max-Planck-Institut für Phys. & Astroph., Führinger Ring 6, 8000 München 40, F.R.G.
N. Bell, Observatoire de Meudon, 92190 Meudon, France
K. Bennett, Space Science Dep. of ESA, ESTEC, Noordwijk, Holland
S. Bensammar, Observatoire de Paris, Meudon, F-92190 Meudon, France
J. Bergeron, ESO-CERN, CH-1211 Geneva 23, Switzerland
A. Bergström, FOA, Sweden
N. Bergvall, Astronomiska Observatoriet, Box 515, S-751 20 Uppsala, Sweden

R. H. Berman, Dept. of Computer Science, Univ. of Reading, Whiteknights, England
C. Bernes, Stockholms Observatorium, 13300 Saltsjöbaden, Sweden
G. Bertelli, Istituto di Astronomia, Vicolo Dell'Osservatorio 5, 35100 Padova, Italy
P. C. Bertiau, Waversebaan 220, 2030 Heverlee Leuven, Belgium
F. Bertola, Osservatorio Astronomico, 35100 Padova, Italy
C. Blanco, CBS. Astrofisica, Citta Univ., 95125 Catania, Italy
L. Bottinelli, Radioastronomie Observatoire de Meudon, 92190 Meudon, France
C. Brihaye, 14 Schemeringlaan, 1900 Overijse, Belgium
B. Byrne, Armagh Observatory, Armagh BT 61 9DG, Ireland
J. Bystedt, Stockholms Observatorium, 13300 Saltsjöbaden, Sweden
A. Caimmi, Istituto di Astronomia, Vicolo dell'Osservatorio 5, 35100 Padova, Italy
D. Callebaut, Univ. Instell. Antw., Dept. Natuurkunde, Universiteitsplein 1, 8-2610 Wilrijk, Belgium
R. Canal, Dep. de Fisica Univ. de Barcelona, Avda GMO Franco, 647, Barcelona, Spain
M. Carlsson, Astronomiska Observatoriet, Box 515, S-751 20 Uppsala, Sweden
A. Cassatella, Satellite Tracking Station, Villafranca del Castillo, P.O. Box 54065 Madrid, Spain
L. M. Celnikier, Observatoire de Meudon, 192 Meudon, France
Yu. E. Charikov, Physio-Technical Institute, U.S.S.R. Academy of Sciences, Leningrad, U.S.S.R.
E. Charvet, Centre d'Etude de Limiel, Villeneuve St. Georges, France
M. T. Chauville, 10 Rue Vandrezanne, 75644 Paris Cedex 13, France
V. Chedia, Abastumani Astrophysical Observatory, Georgian Academy of Sciences, Abastumani, U.S.S.R.
C. Chiuderi, Osservatorio Astrofisico de Arcetri, Largo e Fermi 5, 50125 Firenze, Italy
N. Chugai, Astr. Counc. of the U.S.S.R. Acad. of Sci., Moscow, U.S.S.R.
T. Ciurla, Astr. Obs. of the Wroclaw Univ., Ul Kopernika 11, 51-622 Wroclaw, Poland
J. Clavel, Villafranca del Cast. Sat. Tracking Stn., P.O. Box 54065, Madrid, Spain
J.-P. de Greve, Astrophys. Inst., Vrije Univ. Brussel, Pleinlaan 2, B-1050 Brussels, Belgium
J. Deliyannis, Laboratory of Astronomy-Univ. of Athens, Panepistimiopolis-Aten 621, Greece
Dravins, Lunds Observatorium, Svanegatan 9, 222 24 Lund, Sweden
L. Dunker, Astronomiska Observatoriet, Box 515, S-751 20 Uppsala, Sweden
W. Eichendorf, Astr. Inst. der Ruhr-Universität, Postfach 102148, D-4630 Bochum 1, F.R.G.
Y. Ekedahl, Störtloppsvägen 8 B, S-63357 Eskilstuna, Sweden
A. Ekman, Astronomiska Observatoriet, Box 515, S-751 20 Uppsala, Sweden
A. Elvius, Stockholms Observatorium, 133 00 Saltsjöbaden, Sweden
T. Elvius, Lunds Observatorium, Svaneg. 9, 222 24 Lund, Sweden
L. Erculiani-Abati, Istituto di Astronomia, Vicolo dell'Osservatorio 5,

35100 Padova, Italy
K. Eriksson, Astronomiska Observatoriet, Box 515, S-751 20 Uppsala, Sweden
M. Felli, Osservatorio Astrofisico di Arcetri, Largo e Fermi 5, Firenze, Italy
P. Flin, Observatorium Astronomiczne, Uniwersytet Jagiellonski, Krakow, Ul. Kopernika 27, Poland
C. Fransson, Institutionen för Astronomi, Svanegatan 9, 222 24 Lund, Sweden
G. Gahm, Stockholms Observatorium, 133 00 Saltsjöbaden, Sweden
G. Gaida, Landessternwarte, Königsstuhl, D 6900 Heidelberg, F.R.G.
H.-P. Gail, Inst. für Theor. Astrophys., Im Neuenheimer Feld 294, D-6900 Heidelberg, F.R.G.
P. Gammergård, Astronomisk Institut, Langelandsgade, DK 800 Aarhus, Denmark
D. P. Gilra, Kapteyn Astr. Inst., Univ. of Groningen, P.O. Box 800, 9700 AV Groningen, Holland
Yu. N. Gnedin, Physio-Technical Institute, U.S.S.R. Academy of Sciences, Leningrad, U.S.S.R.
B. F. Gordiets, Physical Institute, U.S.S.R. Academy of Sciences, Moscow, U.S.S.R.
L. Gouguenheim, Radioastronomie Observatoire de Meudon, 92190 Meudon, France
T. Grönningsaeter, Inst. of Theor. Astrophys., Univ. of Oslo, P.O. Box 1029, Blindern, Oslo 3, Norway
J. Mayo Greenberg, Lab. Astrophys., Huyghens Lab., Leiden Univ., Wassenaarseweg 78, Leiden, Holland 2405
G. Guerrero, Osservatorio Astronomico, Via E Bianchi 46, 22055 Merate, Italy
J. Guibert, Radioastronomie Obs. de Meudon, F-92190 Meudon, France
A. L. Gulbudagian, Byurakan Astrophysical Observatory, Armenian Academy of Sciences, Byurakan, U.S.S.R.
B. Gustavsson, Astronomiska Observatoriet, Box 515, S-751, 20 Uppsala Sweden
B. N. G. Guthrie, Royal Observatory, Blackford Hill, Edinburgh EH9 3HJ, England
W. Hagen, Huyghens Laboratorium, Wassenaarseweg 78, Leiden 2405, Holland
M.R.S. Hawkins, Royal Observatory, Blackford Hill, Edinburgh, Scotland
B. Hedin, Astronomiska Observatoriet, Box 515, S-751 20 Uppsala, Sweden
J. Heidmann, Observatoire, 92190 Meudon, France
G. Henriksson, Astronomiska Observatoriet, Box 515, S-751 20 Uppsala, Sweden
J. Heyvaerts, Observatoire de Meudon, D.A.F., 9219 Meudon, France
A. Hjalmarson, Onsala Space Observatory, 430 34 Onsala, Sweden
L. Hultqvist, Stockholms Observatorium, 133 00 Saltsjöbaden, Sweden
H. E. Jörgensen, Astronomisk Observatorium, Östfr. Voldgade 3, DK-1350 Copenhagen, Denmark
L. Johansson, Onsala Space Observatory, 430 34 Onsala, Sweden
P. Joräs, Inst. of Theor. Astrophys., Univ. of Oslo, P.O. Box 1029, Blindern, Oslo 3, Norway
M. P. Kalinkov, Dep. of Astronomy, Bulg. Academy of Sciences

7 Nov. Str. 1, Sofia, Bulgaria
W. H. Kegel, Inst. für Theor. Astrophys., Im Neuenheimer Feld 294, D-6900 Heidelberg, F.R.G.
Kellerman, Max Planck Inst. für Radioastronomie, Auf dem Hügel 69, D-5300 Bonn, F.R.G.
E. Khachikian, Byurakan Astrophysical Observatory, Armenian Academy of Sciences, Byurakan, U.S.S.R.
U. A. Khaud, Inst. for Astrophys. & Phys. of Atmosphere, Estonian Academy of Sciences, Tartu, U.S.S.R.
A. Kinnander, Astronomiska Observatoriet, Box 515, S-751 20 Uppsala, Sweden.
T. A. Kipper, Inst. for Astrophys. & Phys. of Atmosphere, Estonian Academy of Sciences, Tartu, U.S.S.R.
V. S. Kislyuk, Main Astronomical Observatory, Ukrainian Academy of Sciences, Kiev, U.S.S.R.
L. Koch, Centre d'Etude Nucleaires de Saclay, B.P. 2, 91190 Gif-sur-Yvette, France
E. Kontizas, Astr. Inst. Nat. Obs. of Athens, Thission, Aten 306, Greece
T. Korhonen, Astr. Obs. University of Turku, SF-20500 Turku, Finland
K. Kostova, Astronomiska Observatoriet, Box 515, S-751 20 Uppsala, Sweden
J. Krautter, Landessternwarte, 6900 Heidelberg-Königstuhl, F.R.G.
J. M. Kreiner, Spoldzielccw 8 M 52, 30-682 Krakow 47, Poland
J. Krelowski, Inst. of Astr. at Copernicus Univ., Ul Chopina 12-18, PL-87-100 Torun, Poland
C.-I. Lagerkvist, Astronomiska Observatoriet, Box 515, S-751 20 Uppsala, Sweden
M. Laget, Lab. d'Astronomie Spatiale, 13012 Marseille, France
P. Laques, Observatoire du Pic du Midi, 65200 Bagneres de Bigorre, France
G. Larsson-Leander, Institutionen för Astronomi, Svanegatan 9, 222 24 Lund, Sweden
A. Lauberts, Astronomiska Observatoriet, Box 515, S-751 20 Uppsala, Sweden
S. Laustsen, Eso Cern, CH-1211 Geneve, Switzerland
P. Ledoux, Institut d'Astrophysique, Avenue de Cointe 5, 4200 Ougree, Belgium
A. Liljegren, Astronomiska Observatoriet, Box 515, S-751 20 Uppsala, Sweden
P. Lindblad, Stockholms Observatorium, 13300 Saltsjöbaden, Sweden
L. Lindegren, Lunds Observatorium, Svanegatan 9, 22 24 Lund, Sweden
H. Lindgren, Institutionen för Astronomi, Svanegatan 9, 222 24 Lund, Sweden
R. Liseau, Stockholms Observatorium, 133 00 Saltsjöbaden, Sweden
L. S. Lubinkov, Crimean Astrophysical Observatory, U.S.S.R. Academy of Sciences, Crimea, U.S.S.R.
K. Lundgren, Astronomiska Observatoriet, Box 515, S-751 20 Uppsala, Sweden
L. Lundin, Astronomiska Observatoriet, Box 515, S-751 20 Uppsala, Sweden
I. Lundström, Inst. för Astronomi, Svanegatan 9, S-22224 Lund, Sweden
G. Lyngä, Lunds Observatorium, 22224 Lund, Sweden

W. J. Maciel, Kapteyn Astronomical Observatory, P.O. Box 800, Groningen, Holland
L. Malkamäki, Univ. of Helsinki Obs. and Astrophys. Lab., Tähtitorninmäki, SF-00130 Helsinki, Finland
M. Malmort, Skyttev. 18 c 133 00 Saltsjöbaden, Sweden
P. Maltby, Institutt for Teoretisk Astrofysikk, Postboks 1029, Blindern Oslo, Norway
L. Mantegazza, Osservatorio Astronomico, 22055 Merate (Como), Italy
T. Markkanen, Univ. of Helsinki Obs. and Astrophys. Lab., Tähtitorinmäki, SF-00130 Helsinki 13, Finland
E. Maurice, Observatoire de Marseille, 2, Place Le Verrier, F-13004 Marseille, France
M. C. McCarthy, Specola Vaticana, Vatican City State, Italy
Yu. G. Melik-Alaverdian, Byurakan Astrophysical Observatory, Armenian Academy of Sciences, Byurakan, U.S.S.R.
M. D. Metreveli, Abastumani Astrophysical Observatory, Georgian Academy of Sciences, Abastumani, U.S.S.R.
R. J. Mitchell, Mullard Space Science Lab., Holmbury St. Mary, Dorking, Surrey, England
I. G. Mitrofanov, Physio-Technical Institute, U.S.S.R. Academy of Sciences, Leningrad, U.S.S.R.
T. Ch. Mouschovias, Coll. of Liberal Arts and Sciences, Dept. of Astronomy, Obs., Urbana, Ill. 61801, U.S.A.
A. Natta, Laboratorio Astrofisica Spaziale, Casella Postale 67, 00044 Frascati, Italy
R. Nesci, Lab. Astrofisica Spatiale, CP 67, 00044 Frascati, Italy
B. Nordström, Copenhagen University Observatory, Brorfelde, DK-4340 Töllöse, Denmark
A. Norman, Huygens Laboratory, Wassenaarseweg 78, Leiden, Holland
F. Occhionero, Laboratorio de Astrofisica Spaziale, Casella Postale 67, 00044 Frascati, Italy
T. Oja, Astronomiska Observatoriet, Box 515, S-751 20 Uppsala, Sweden
N. Olander, Astronomiska Observatoriet, Box 515, S-751 20 Uppsala, Sweden
F. Pacini, FSO, C/O Cern, 1211 Geneve, Switzerland
B.E.J. Pagel, Royal Greenwich Obs., Herstmonceux, Sussex, England
P. L. Palmer, Inst. of Astronomy, The Observatories, Madingley Road, Cambridge, OB3 OHA, England
N. Panagia, Laboratorio di Radioastronomia, Via Irnerio 46, 40126 Bologna, Italy
V. E. Panchuk, Special Astrophysical Observatory, U.S.S.R. Academy of Sciences, Caucasus, Nizhniy, Arkhyz, U.S.S.R.
V. Pankonin, Max-Planck-Institut für Radioastronomie, Auf dem Hügel 69, D-5300 Bonn, F.R.G.
P. Paolicchi, Osservatorio Astronomico, 22055 Merate, Italy
J.-C. Pecker, Institute d'Astrophysique, 98 bis Bd. Arage, 75014 Paris, France
M. Petrou, Inst. of Astronomy, The Observatories, Madingley Road, Cambridge CB3 OHA, England
P. Petrov, Stockholms Observatorium, 13300 Saltsjöbaden, Sweden
B. Pettersen, Inst. för Teor. Astrophysikk., Univ. i. Oslo,

PB 1029, Blindern, Oslo 3, Norway
B. Petterson, Astronomiska Observatoriet, Box 515, S-751 20 Uppsala, Sweden
W. Pietsch, Max-Planck-Institut, 8046 Garching, F.R.G.
A. Pinotsis, Lab. of Astro. Panepistimiopolis, Ilisa 621, Aten, Grecce
M. Popova, Dept. of Astronomy Bulg. Academy of Sciences, 7 Nov. Str. 1, Sofia, Bulgaria
V. I. Pronik, Crimean Astrophysical Observatory, Crimea, U.S.S.R.
S. Röser, MPI für Kernphysik, Postfach 103980, D-6900 Heidelberg, F.R.G.
M. Raoult, Centre d'Etude Nucleaires de Sacly, B.P. 2, 91190 Gif-sur-Yvette, France
A. Renzini, Osservatorio Astronomico, Casella Postale 596, 40100 Bologna, Italy
N. Roos, Huygens Laboratorium, Wassenaarseweg 78, Leiden 2405, Holland
M. Rosa, Landessternwarte, Königsstuhl, D 6900 Heidelberg, F.R.G.
K. Rudnicki, Hugelsweg 39, CH 4143 Dornach, Switzerland
A. A. Ruzmaikin, Institute for Applied Mathematics, U.S.S.R. Academy of Sciences, Moscow, U.S.S.R.
G. Rydbeck, Onsala Space Observatory, 430 34 Onsala, Sweden
M. Salvati, Laboratorio di Astrofisica Spaziale, Casella Postale 67, 00044 Frascati, Italy
K. Särg, Inst. för Astronomi, Svaneg. 9, 222 24 Lund, Sweden
A. Sandquist, Stockholms Observatorium, 133 00 Saltsjöbaden, Sweden
R. A. Sarkisian, Byurakan Astrophysical Observatory, Armenian Academy of Sciences, Byurakan, U.S.S.R.
M. Saxner, Astronomiska Observatoriet, Box 515, S-751 20 Uppsala, Sweden
C. Schalen, Lunds Observatorium, Svaneg. 9, 222 24 Lund, Sweden
D. Schallwich, Astronomisches Inst. der Ruhr-Univ., Postfach 102148, D-4630 Bochum 1, F.R.G.
W. Schlosser, Astronomisches Institut, Postfach 102148, D-4630 Bochum 1, F.R.G.
K.-H. Schmidt, Zentral Institut für Astrophys., Rosa-Luxemburg-Strasse 17a, 1502 Potsdam-Babelsberg, G.D.R.
G. F. O. Schnur, European Southern Observatory, Casilla 16317, Correo 9, Santiago, Chile
E. Sedlmayr, Inst. für Theor. Astrophys., Im Neuenheimer Feld 294, D-6900 Heidelberg, F.R.G.
G. Sedmak, Osservatorio Astronomico, Via G. B. 7, Tiepolo 11, 34131 Trieste, Italy
J. Sellwood, Stockholms Observatorium, 133 00 Saltsjöbaden, Sweden
W. W. Shane, Huygens Laboratorium, Wassenaarseweg 78, 2300 RA Leiden, Holland
G. B. Sholomitsky, Institute of Space Research, U.S.S.R. Academy of Sciences, Moscow, U.S.S.R.
J. Sikorski, Inst. of Phys. Univ. of Gdansk, Ul Wita Stwosza 57, 80-952 Gdansk, Poland
F. Simien, Observatoire de Marseille, 2, Place le Verrier, 13004 Marseille, France
S. Sivertsen, IMR Universitetet i Tromsö, P.O. Box 953, 9001 Tromsö, Norway

O. Soldal, Inst. of Theor. Astrophys., Univ. of Oslo, P.O. Box 1029, Blindern, Oslo 3, Norway

J. Solheim, Nordlysobs Univ. i. Tromsö, 9001 Tromsö, Norway

B. V. Somov, Physical Institute, U.S.S.R. Academy of Sciences, Moscow, U.S.S.R.

R. Staubert, Astronomisches Institut, Waldhauser Str. 64, D 74 Tirbingen, F.R.G.

M. Steinbach, Astro. Dept. Wea, Veb. Carl Zeiss Jena, 69 Jena, Ötyskland

B. Stenholm, Inst. för Astronomi, Svanegatan 9, 22224 Lund, Sweden

L. G. Stenholm, Inst. für Theor. Astrophysik, Heidelberg, Im Nevenheimer Feld 294, 6900 Heidelberg, F.R.G.

A. Strobel, Instytut Astronomii UMK, Chopina 12/18, 87-100 Torun, Poland

I. T. Sudzhus, Physical Institute of the Lithuanian Academy of Sciences, Vilnius, U.S.S.R.

N. Svolopoulos, Odos Vassileos Alexandrou 28, Amaroussion-Pefki, Athens, Greece

B. N. Swanenburg, Cosmic Ray Working Group, Wassenaarseweg 78, Leiden, Holland

P. Teerikorpi, Observatory and Astrophysics Lab. Univ., Tähtitorninmäki, SF-00130 Helsinki, Finland

G. Tendrio-Tagle, Max-Planck-Institut für Phys. & Astroph., Föhringer Ring 6, 8000 München 40, F.R.G.

S. Theodossiou, Lab. of Astro., Panepistimiopolis, Ilisia 621, Athens, Greece

F. Torcel, Centre d'Etude de Limiel, B.P. 27, 94190 Villeneuve St. Georges, France

T. Toro, Dep. of Phys. Univ. of Timisoara, 1900 Timisoara, Rumania

J. Trumper, Max-Planck-Institut, 8046 Garching, F.R.G.

E. Trussoni, Laboratorio di Cosmo-Geofisica del CNA, Corso M., D'Azeglio 46, I-10125 Torino, Italy

Maria Tsoga, Laboratory of Astronomy Univ. of Athens, Panepistimiopolis-Aten, Greece

C. van der Bijlt, Laboratory Astrophysica, Huygens Lab., Wassenaarseweg 78, Leiden 2405, Holland

J.J.A.M. van der Mullen, P. Windhausenweg 25, Breda, Holland

F. van Leeuwen, Huygens Laboratorium, Wassenaarseweg 78, Leiden 2405. Holland

D.A. Varshalovich, Physio-Technical Institute, U.S.S.R. Academy of Sciences, Leningrad, U.S.S.R.

G. Vettolani, Laboratorio di Radioastron. CNR, Via Irnerio 46, Bologna, Italy

J.-L. Vidal, Observatoire de Pic du Midi, 65200 Bagnères de Bigorre, France

T. Vieira, Astronomiska Observatoriet, Box 515, S-751 20 Uppsala, Sweden

F. Vikanes, Inst. of Theor. Astrophys., Univ. of Oslo, P.O. Box 1029, Blindern, Oslo 3, Norway

W. Voges, Max-Planck-Institut, D 8046 Garching, F.R.G.

A. Wallenquist, Astronomiska Observatoriet, Box 515, S-751 20 Uppsala, Sweden

C. M. Walmsley, Max Planck Institut für Radioastr., Auf dem Hügel 69, 5300 Bonn, F.R.G.
V. Weidemann, Inst. für Theor. Physik und Sternwarte, Olshausenstrasse, 2300 Kiel 1, F.R.G.
G. Welin, Astronomiska Observatoriet, Box 515, S-751 20 Uppsala, Sweden
P. R. Wesselius, Dept. of Space Research, P.O. Box 800, Groningen, Holland
B. Westerlund, Astronomiska Observatoriet, Box 515, S-751 20 Uppsala, Sweden
R. Wielen, Astronomisches Rechen-Institut, Mönchhofstrasse 12-14, D-6900 Heidelberg 1, F.R.G.
A. Wiesmeier, Astronomisches Institut, Postfach 102148, D-4630 Bochum 1, F.R.G.
A. Winnberg, Max-Planck-Institut für Radioastronomie, Auf dem Hügel 69, D-5300 Bonn, F.R.G.
S. Wramdemark, Institutionen för Astronomi, Svanegatan 9, 222 24 Lund, Sweden
Ya. A. Yaaniste, Inst. for Astrophys. & Phys. of Atmosphere, Estonian Academy of Sciences, Tartu, U.S.S.R.
N. Zentelis, Svartviksringen 4, 133 00 Saltsjöbaden, Sweden
H. Zimmermann, Max-Planck-Institut, 8046 Garching, F.R.G.
J. Ziznovsky, Astr. Inst. Slovak Acad. of Sciences, 05960 Tatranska Lomnica, Czechoslovakia

GAMMA-RAY SOURCES: A PUZZLE, OR A PIECE OF THE PUZZLE?

B.N. Swanenburg
Cosmic Ray Working Group, Huygens Laboratory,
Wassenaarseweg 78, 2300 RA Leiden, The Netherlands

ABSTRACT

The gamma-ray satellite COS-B has, at this time, detected twenty high-energy (>50 MeV) gamma-ray sources. Two sources are pulsars (PSR0531+21 and PSR0833-45) and one source has been identified with the quasar 3C273. The remaining sources are as yet unidentified galactic objects. Compact objects in which acceleration of particles and emission of gamma rays occur in a well ordered geometry (like pulsars) seem to be attractive candidates. The analogy with the well defined geometry of quasars and radio galaxies suggests that their nucleus contains a powerful gamma-ray source.

INTRODUCTION

The subject of gamma-ray astronomy was introduced a long time ago. In the early 50's it was realised that nuclear and electromagnetic interactions of cosmic rays with interstellar matter would produce measurable fluxes of high-energy (>50 MeV) gamma radiation along the galactic equator (Hayakawa 1952; Hutchinson 1952, and Morrison 1958). Clearly, measurements of the spatial and spectral distribution of galactic gamma rays would provide unique information regarding galactic structure, specifically with respect to the distribution of gas and cosmic rays. Difficulties related to the expected low fluxes, high background levels and the interfering effects of the earth atmosphere have delayed experimental progress for many years. Not until 1967 was the gamma-ray galaxy mapped in some detail by the pioneering experiment on OSO-3 (Kraushaar *et al.* 1972).

In the same timespan astronomy in general evolved enormously. The discovery of powerful radio galaxies, quasars, X-ray sources and pulsars made us aware of the unexpected energetics of many, very different, types of astronomical objects. Such objects could be the ideal site for the production of gamma rays. Alternatively, since gamma rays can only be produced by very energetic particles

Bengt E. Westerlund (ed.), Stars and Star Systems, 1–13.

in the presence of matter or fields (or by annihilation of matter with antimatter), detection of gamma rays from particular objects demonstrates the existence of powerful non-thermal processes. The detection of gamma rays could reveal the *real* energetics of the source. The sobering fact that firm predictions for *detectable* fluxes from known objects were lacking, did not diminish the enthusiasm with which the development of more sensitive instrumentation with better angular resolution was undertaken. Early investigations were carried out using balloon platforms (for a review, see Fazio 1973), but the long exposures available with satellites were required before firm results could emerge.

The first second-generation gamma-ray observatory, SAS-2, was launched in 1972. This experiment established beyond question the gamma-ray emission from the Crab and Vela pulsars and discovered an as yet unidentified gamma-ray source. Unfortunately the mission failed after only seven months. Consequently, only part of the sky was covered and intriguing hints for gamma-ray emission from two other pulsars and from Cyg-X3 could not be rechecked. An overview of most of the results of this mission is given by Fichtel (1977).

On August 9, 1975, the ESA gamma-ray observatory COS-B was launched. The experiment was developed by a collaboration of six institutes, known as the 'Caravane' collaboration. The prolonged and still ongoing observations result in by far the most sensitive investigation of galactic and extra-galactic gamma radiation to date. Apart from new data on the general galactic emission (Bennett *et al.* 1977a; Paul *et al.* 1978), the results introduce a new astronomical puzzle, *viz.* the discovery of numerous high-energy gamma-ray sources. The situation reminds of the early days of radio and X-ray astronomy. New objects are found which do not fit within existing concepts.

It is premature to present a balanced review on this subject. Results obtained with the COS-B satellite are continuously presenting new facts. By no means has any consensus been reached regarding the nature of the newly discovered gamma-ray sources. The aim of this paper is to present a status report of the experimental facts. Boundary conditions applying to the formulation of models for gamma-ray sources will be sketched. These considerations lead to the suggestion that gamma-rays play a dominant role in the transport of energy in quasars and radio galaxies.

The experimental data which form the bases for this paper are presented on behalf of the 'Caravane' collaboration. The subsequent discussions are the responsibility of the author.

INSTRUMENTATION

The combination of the physical processes involved in the detection of gamma rays, the low inherent flux levels, and the high cosmic-ray background has led to the conception of gamma-ray detectors of which the characteristics are very different from those of observatories working at any other wavelengths. Detailed descriptions of the COS-B instrument, its calibration and the operations, have been published (Bignami *et al.* 1975; Christ *et al.* 1974; Scarsi *et al.* 1977). Some typical characteristics are summarized in table 1.

Table 1. Typical characteristics of COS-B.

Energy range	50 MeV - 10 GeV
Wavelength range	3×10^{-12}-1.5×10^{-14} cm
Frequency range	10^{22} -2×10^{24} Hz
Energy resolution	50-100 % (FWHM)
Sensitive area	5 - 50 cm^2
Angular resolution	1.5 - 10^{o}(FWHM)
Field of view	$\sim 40^{o}$ (full cone)
Duration of one observation	4 - 6 weeks
Number of photons collected	strongest source ∿30 } per day weakest source ∿ 1 } per day
Weight of instrument	120 kg

A few points are emphasized. The instrument covers a very large range in energy which, in combination with the moderate energy resolution, permits a good asessment of energy spectra. It is noted that no sharp spectral features are expected at these energies. Of particular importance is the need for prolonged observations if meaningful results are to be obtained. Even then a very limited number of photons are collected from a typical gamma-ray source, making a search for periodic or irregular variability virtually impossible, except in cases where the timing characteristics from the source are accurately known from other observations. The angular resolution, dictated by the physics of the detection process, sets severe limits on the obtainable positional accuracy for sources. These basic limitations are compensated to some extent by the large field of view of the instrument, which permits that several sources or candidate sources, can be studied simultaneously.

OBSERVATIONAL RESULTS

To date COS-B has completed the survey of the galactic plane effectively to latitudes up to 15-20°. Several regions have been observed more than once as a result of overlap between adjacent observations or because repeated observations were considered essential. In addition, quite some time has been devoted to high latitude observations. All data have been inspected for big surprises; however, normally conclusions are only firm after indepth analyses of the data, which involves an ongoing learning process. The results presented here are in the form of a status report. Further detailed analyses may change the picture slightly.

Figure 1 illustrates the distribution of 20 gamma-ray sources in a projection of the sky in galactic coordinates. The positions are listed in table 2. A gamma-ray source is defined as a localised, statistically significant, enhancement of the observed flux, consistent with the expected distribution of a true point source. In general this implies that the true angular extent of the source is 2° or less. The majority of these sources have been looked at more than once and have been *seen* more than once. Statistical estimates indicate that with high probability all observed sources are real. But the eventuality that one or two are in fact chance fluctuations of the background radiation can never be ruled out with absolute certainty.

If we accept the data as it stands, one conclusion is imminent: gamma-ray sources are galactic. Up to now only one source has been detected at a high galactic latitude.

The fact that the gamma-ray sources are so closely aligned with the galactic equator makes it very difficult in certain regions to resolve them uniquely from the underlying more or less diffuse galactic emission. At present those regions which do contain *significant* structure but which have yet to be resolved are indicated by rectangles in figure 1. In some cases, outstanding sources in these complex regions are resolved already, as shown in the figure. Further data analysis and additional observations will definitely lead to the clear recognition of a number of sources in these complex regions.

Only three of the twenty gamma-ray sources have been identified with known astronomical objects. Detailed characteristics of the gamma-ray pulsars PSR0531+21 and PSR0833-45 have been published (Bennet *et al.* 1977b and Buccheri *et al.* 1978). These identifications are unquestionable because of the observed gamma-ray pulsations at the exact frequency of the radio pulsations. The source at high galactic latitude has been identified with the quasar 3C273 (Swanenburg *et al.* 1978).

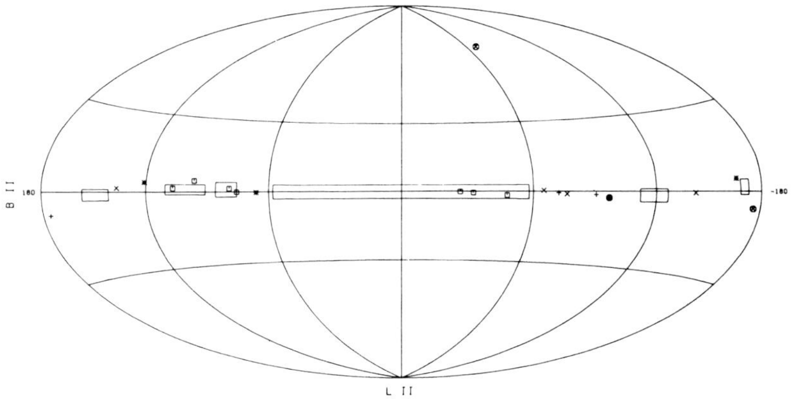

Figure 1. Distribution of gamma-ray sources. Symbols used for spectral classification (see text): ✱ hard, x medium, + soft, ◻ no class. Identified sources are indicated by a circle. Rectangles: complex regions.

Thirteen unidentified sources were investigated in two energy ranges. A spectral class has been assigned according to the ratio of counts recorded between 50 and 150 MeV and above 150 MeV. A large value for this ratio indicates a source with a 'soft' spectrum, a low value indicates a 'hard' spectrum. Sources with intermediate values are called 'medium'. In figure 1 the spectral classification of the sources is indicated. In this classification the Vela pulsar (PSR0833-45) has a hard spectrum, whereas the Crab pulsar (PSR0531+21) has a medium spectrum. The observed ratios between the number of counts at low energy and those at high energy vary by as much as a factor of five. This indicates that the various sources exhibit very different energy spectra. Spectral analysis of sources in or near complex regions has not been possible yet, because the low energy data is heavily confused.

Apart from the periodic variable sources, the gamma-ray pulsars, variability on a timescale between 1 month and 1 year has been observed in several cases. This could be concluded from the comparison of data of the same region of the sky obtained in different observation periods. Unfortunately, this type of variability has up-to-now only been unambiguously observed in the so-called complex regions, so that the variability can not yet be associated with specific sources.

Taking 70 pc as the minimum scale-height of any galactic population, the observed average latitude of the sources ($\sim 2^o$) yields a value of 2 kpc as a minimum typical distance for the sources. The energy flux (>100 MeV) ranges from 3 to 15 times 10^{-10} erg $cm^{-2}s^{-1}$.

Table 2. Gamma-ray sources detected by COS-B. (see notes below)

Galactic coordinates		Remarks
l^{II}	b^{II}	
65.5	-0.0	CG64+0
75.0	+0.0	CG75+0
78.5	+1.5	CG78+1
95.5	+4.5	
106.0	+1.5	
121.0	+3.5	CG121+3
135.5	+1.5	CG135+1
176.0	-7.0	CG176-7
184.5	-5.5	CG185-5, PSR0531+21
195.5	+4.5	CG195+4
219.0	-0.5	
263.5	-2.5	CG263-2, PSR0833-45
270.0	-1.0	
284.0	-1.0	
288.5	-0.5	
291.0	+65.0	CG291+65, 3C273
295.5	+0.5	
312.0	-1.5	CG312-1
327.5	-0.5	CG327-0
333.5	+0.0	CG333+0

Notes.

1. For 'CG' sources see Hermsen *et al.* (1977).
2. The information is preliminary; analysis continues.
3. Error circles for positions have radii between 0.5^{0} and 2.0^{0}.
4. With the exception of CG263-2 source intensities (>100 MeV) are within the range 0.2 to 1.0 times that of CG185-5.

If the sources emit isotropically their typical luminosity will, therefore, be in the range of 10^{35}-10^{36} erg s^{-1}.

SEARCH FOR COUNTERPARTS

Soon after the presentation of the first 10 new gamma-ray sources (Hermsen *et al.* 1977) many suggestions for associations with known objects were put forward (e.g. Abdulwahab and Morrison 1978; Gregory and Taylor 1978; Julien and Helmken 1978; Lamb 1978; Maraschi *et al.* 1978). None of these searches have led to unambiguous identifications.

It is stressed that firm identification of a particular gamma-ray source on the basis of positional coincidence alone is never possible, in view of the large area of the error boxes (1-10 sq degr). Identification may follow from a particular behaviour (e.g. pulsars) or may be attempted on a sample bases by correlation against a selected set of objects.

The list of gamma-ray sources as given in table 2 was checked for correlations against catalogues of X-ray sources (Forman *et al.* 1978), pulsars (Taylor and Manchester 1975) and supernova remnants (Ilovaisky and Lequeux 1972). This search leads to the following conclusions:

- Depending on how literally we take the gamma-ray source positions one finds three to six possible associations with weak (less than 10 UHURU counts) X-ray sources. Even if not all of the associations are coincidences by chance, it follows that the X-ray luminosity of the gamma-ray sources is less than 20 % of the gamma-ray luminosity. For the majority of gamma-ray sources not even a candidate X-ray counterpart has been found. On the other hand, if some of the associations are real, we may conclude that X-ray astronomy is reaching the sensitivity required to see the counterparts of gamma-ray sources.
- Because none of the gamma-ray sources are associated with a very strong radio source it may be safely concluded that the radio luminosity is very much less than the gamma-ray luminosity.
- The situation with respect to supernova remnants is interesting. First, it is noted that the mean galactic latitude of supernova remnants is about equal to the mean latitude of gamma-ray sources. In addition, although no supernova remnants have been detected within the error box of 11 gamma-ray sources, it is found that in one case the gamma-ray error box contains one supernova remnant and in five cases a gamma-ray error box contains two or more supernova remnants. Those findings suggest that regions of the Galaxy which favour the occurence of supernovae provide a convenient environment for gamma-ray sources to exist.
- Apart from the Crab and Vela pulsars none of the sources listed in table 2 have been identified with a known pulsar.

For the understanding of gamma-ray sources the identification of at least some of them with objects which can be studied in detail at X-ray, optical and radio wavelengths is of utmost importance. The discussion above illustrates the type of difficulties which will be encountered. With some luck a brakethrough may come from the HEAO-B satellite, which will be able to provide very accurate positions for X-ray sources below current detection limits. If not too many X-ray sources are found within each gamma-ray error box, firm identifications may follow.

SOURCE MODELS

The various processes leading to the production of high energy gamma rays have been studied in depth (e.g. Chupp 1976, and references therein). The fundamental ingredients for the production of gamma-rays are, on the one hand, high energy particles, and on the other hand a target, consisting of either matter or electromagnetic fields. Gamma-ray source models should combine these ingredients in such a way that gamma-ray sources with the properties as summarized in table 3 are predicted.

Table 3. Characteristics of galactic gamma-ray sources.

Angular size	$\leq 2^{o}$
Intensity (>100 MeV)	$\sim$ (0.7-4) x 10^{-6} ph cm^{-2} s^{-1}
Energy flux(>100 MeV)	$\sim$ (3-15) x 10^{-10} erg cm^{-2} s^{-1}
Different spectral classes	L/H$\sim$0.5-3
At least some sources are variable	
Typical distance	$\geq$ 2 kpc
Typical luminosity	10^{35}- 10^{36} erg s^{-1}
Number	$\geq$ 15 within $60^{o} < l^{II} < 300^{o}$
From lack of identification	$L_x < 0.2 L_\gamma$
	$L_{radio} << L_\gamma$

Several models for selected gamma-ray sources have been proposed (e.g. Abdulwahab and Morrison 1978; Maraschi and Treves 1977a; Strong 1977, Davies *et al.* 1978). The models do not at all converge to a clear picture of what a gamma-ray source will look like. Rather than trying to explain a small sample of the gamma-ray sources, it may be instructive to try to derive boundary conditions for source models assuming that the majority of the gamma-ray sources belong to one class of objects. Such an approach seems justified on the bases that all unidentified gamma-ray sources have similar characteristics. A first attempt along these lines has been made by the author (Swanenburg 1977). Several general considerations can best be discussed with reference to a first order summary of source models as given in table 4.

Table 4. Source models

Type	Extended	Compact	
Energy source	Cosmic rays	Gravitation	'Rotation'
Target	I.S. gas	Plasma, surface	Magnetic field
Example	I.S. cloud	X-ray source (NS or BH)	Pulsar
Character	Chaotic	Chaotic	Ordered
Advantage	No X-rays	Ample energy	It works
Problems	Not observed Variability Spectra	$L_X >> L_\gamma$	Energetics of NS marginal

Obviously a significant distinction exists between extended and compact sources. The most natural candidate for extended sources is abundantly found in the Galaxy in the form of interstellar clouds bombarded by energetic particles, e.g. cosmic rays. (Black and Fazio 1973). Such sources would have little or no X-ray emission. However, clouds in excess of $10^6 M_\odot$ are required to explain the observed gamma-ray intensities, if the cosmic-ray intensity within the clouds is about equal to the local cosmic-ray intensity. Such massive clouds have not been detected in the direction of the gamma-ray sources. In addition little differences in the spectra are expected and the sources could never show time variability. The possibility that less massive clouds with a strong cosmic-ray source embedded inside, produce the observed fluxes can not be ruled out, although it would be expected that such objects would have drawn our attention from radio observations.

It is now well established that compact objects like neutron stars and possibly black holes can generate very energetic phenomena in their surroundings. Energy may be generated as a result of gravitation, like in accreting X-ray sources, or as a result of the loss of rotational energy, as in the case of pulsars.

Accretion models (e.g. Maraschi and Treves 1977b) are attractive because energy is available in unlimited quantities as long as there is sufficient accreting material. The problem is that all models predict a much larger X-ray luminosity than gamma-ray luminosity. Basically, this is due to the fact that the gravitational forces can not accelerate particles to sufficiently high energy for the production of gamma radiation. Plasmas containing turbulent magnetic fields are invoked to provide additional acceleration processes. In this way gamma rays around 100 MeV can indeed be produced. If such processes were to work in the gamma-ray sources, one has to postulate an additional absorption mechanism for the X-rays which are simultaneously produced. This could of course happen in certain cases, but then it remains unexplained why in *all* accretion systems which do produce gamma-rays, the X-rays are heavily absorbed. In addition, one has to consider an other important effect. *Gamma-rays* may be absorbed by photon-photon collisions leading to the production of electron pairs. (Gould and Schréder 1967). It turns out that the mean free path for say 100 MeV gamma-rays in a turbulent plasma with high X-ray photon densities may be as short as a few centimeters. Although this effect needs more quantitative assessment, it may turn out that if gamma-rays are produced via any turbulent (chaotic) acceleration process, the unavoidable production of X-rays may prohibit the escape of the gamma-rays.

In contrast to accretion models, the acceleration of particles and the generation of gamma-rays in pulsars occurs in a very orderly way. The details of the processes leading to the production of gamma-rays are not known, but the detection of gamma-rays from the Crab and Vela pulsars proves that the system works. The magnetic field of the neutron star imposes a strong geometric ordering in the acceleration and emission processes. The observations also confirm that indeed a major part of the emission occurs at gamma-ray energies. Yet two main problems must be solved before we can accept that the majority of gamma-ray sources are pulsars. First, it is noted that the latitude distribution of pulsars is significantly wider than that of gamma-ray sources. Possible observational selection effects may cause this discrepancy and should be investigated further. Independently the apparent discrepancy may be resolved if only those pulsars which are situated in an environment of relatively high gas densities are efficient gamma-ray producers. A second problem seems more difficult to resolve. Considering that the available energy reflects the loss of rotational energy, which can be directly estimated from the rate of slowing down of pulsars, it turns out that ot the order of 10 per cent or more of the rotational energy must be transformed into high-energy gamma radiation in order to explain the number of gamma-ray sources and their observed luminosities (Swanenburg 1977). For comparison it is noted that the 'efficiency' of the Crab pulsar is less than 10^{-3}.

Theoretical investigations of the production of gamma rays in pulsars is urgently required, but of equal importance is a very sensitive search for pulsars at the location of the gamma-ray sources. Independent of the outcome of these studies the result will be extremely interesting. If it is found that gamma-ray sources are pulsars, we have discovered very efficient gamma-ray machines. On the other hand, if it is established that pulsars in general are not sufficiently energetic to explain the gamma-ray sources, we have to postulate the existence of objects which are more powerful but which preferably share the ordered emission mechanism with pulsars. In this latter case one may have to think in terms of black holes, which may provide the combination of a huge energy reservoir, in the form of angular momentum, and a strong magnetic field (e.g. Ruffini 1972).

At this stage it is of interest to note that the quasar 3C273 has the same general property as the other gamma-ray sources in that the gamma-ray luminosity exceeds the luminosity at other wavelengths. This suggests that considerations applicable to the galactic gamma-ray sources may impose similar boundary conditions for models relevant to the gamma-ray emission of 3C273. If this similarity in appearance is indeed indicative of a similarity in source properties, a strong constraint on source models would be the direct consequence. In this context it is noted that source models with a black hole as key element may more easily be scaled over many orders of magnitude in power, while preserving relatively constant characteristics, than any other model.

DISCUSSION

The newly discovered gamma-ray sources present a new astronomical puzzle. Speculations with respect to the nature of these objects, as discussed in the previous chapter, have to be checked more quantatively against detailed theoretical investigations and against future observations, in particular at complementary wavelength. Irrespective of future developments, one conclusion is eminent: objects of moderate to very high luminosities, and which emit preferentially at gamma-ray energies, do exist. Rather than explaining the gamma-ray emission as the high-energy tail of the emission process, one may have to think in terms of processes where the gamma-rays are produced at an early stage and in which X-ray, optical and radio emission reflect secondary processes. The suggestion that gamma-ray emission is more easily realised in a system governed by a well-defined geometry, as given for instance by a magnetic dipole, raises an intriguing question, *viz.*: is the well-defined geometry as observed in many radio galaxies and quasars in any way related to the problem of gamma-ray sources? In a qualitative sense this question can be answered positively, by assuming that radio galaxies and quasars contain in their nucleus a gamma-ray source emitting beams of very energetic gamma-rays (Swanenburg 1978). These gamma-rays may interact with photons of the microwave background, thereby producing electron pairs. This process provides an efficient mechanism for the transport of energy from the nucleus to the outer radio lobes. If this model proves to be true, gamma-ray sources have been studied already for many years by radio telescopes. In this picture, the discovery of much weaker gamma-ray sources in the Galaxy, brings some of the outstanding questions of radio-galaxies and quasars closer to a solution. However, a new question is raised: how does nature generate gamma-rays secretly?

ACKNOWLEDGEMENT

I thank the Caravane collaboration for their permission to present unpublished data.

REFERENCES

Abdulwahab, M., and Morrison, P.: 1978, Astrophys. J. 221, L33.
Bennett, K., Bignami, G.F., Buccheri, R., Hermsen, H., Kanbach, G., Lebrun, F., Mayer-Hasselwander, H.A., Paul, J.A., Piccinotti, G., Scarsi, L., Soroka, F., Swanenburg, B.N., and Wills, R.D.: 1977a, Proc. 12th ESLAB Symp., ed. Battrick, B., ESTEC Noordwijk, ESA SP 124, 83.
Bennett, K., Bignami, G.F., Boella, G., Buccheri, R., Hermsen, W., Kanbach, G., Lichti,G.G., Masnou, J.L., Mayer-Hasselwander, H.A., Paul, J.A., Scarsi, L., Swanenburg, B.N., Taylor, B.G., and Wills, R.D.: 1977b, Astron. Astrophys. 61, 279.
Bignami, G.F., Boella, G., Burger, J.J., Keirle, P., Mayer-Hasselwander, H.A., Paul, J.A., Pfeffermann, E., Scarsi, L., Swanenburg, B.N., Taylor, B.G., Voges, W., and Wills, R.D.: 1975, Space Sci. Instrum. 1, 245.
Black, J.M., and Fazio, G.G. : 1973, Astrophys. J. 185, L7.
Buccheri, R., Caraveo, P., D'Amico, N., Hermsen, W., Kanbach, G., Lichti, G.G., Masnou, J.L., Wills, R.D., Manchester, R.N., and Newton, L.M.: 1978, Astron. Astrophys. 69, 141.
Christ. H., Peters, F., Bignami, G.F., Burger, J.J., Hermsen, W., Paul, J.A., Pfeffermann, E., Taylor, B.G., Voges, W., and Wills, R.D.: 1974, Nucl. Instrum. and Methods 116, 477.
Chupp, E.L.: 1976, *Gamma Ray Astronomy*, Geophysics and Astrophysics Monographs 14, D. Reidel Publ. Co., Dordrecht, Holland
Davies, R.E., Fabian, A.C., and Pringle, J.E.: 1978, Nature 271, 634.
Fazio, G.G.: 1973, *X- and Gamma-Ray Astronomy* p. 303, Eds. H. Bradt and R. Giacconi, D. Reidel, Dordrecht, Holland
Fichtel, C.E.: 1977, Space Sci Rev. 20, 191.
Forman, W., Jones, C., Cominsky, L., Julien, P., Murray, S., Peters, G., Tananbaum, H., and Giacconi, R.: 1978, Astrophys. J. Suppl. Ser., in press.
Gould, R.J., and Schréder, G.P.: 1967, Phys. Review 155, 1404.
Gregory, P.C., and Taylor, A.R.: 1978, Nature 272, 704.
Hayakawa, S. : 1952, Progr. Theor. Phys. 8, 571.
Hermsen, W., Swanenburg, B.N., Bignami, G.F., Boella, G., Buccheri, R., Scarsi, L., Kanbach, G., Mayer-Hasselwander, H.A., Masnou, J.L., Paul, J.A., Bennett, K., Higdon, J.C., Lichti, G.G., Taylor, B.G., and Wills, R.D.: 1977, Nature 269, 494.
Hutchinson, G.W.: 1952, Phil. Mag. 43, 847.
Ilovaisky, S.A., and Lequeux, J.: 1972, Astron. Astrophys. 18, 169.
Julien, P.F. and Helmken, H.F.: 1978, Nature 272, 699.
Kraushaar, W.L., Clark, G.W., Garmire, G.P., Borken, R., Higbie, P., Leong, V. and Thorsos, T.: 1972, Astrophys. J. 177, 341.

Lamb, R.C.: 1978, Nature 272, 429.
Maraschi, L., Markert, T., Apparao, K.M.V., Bradt, H., Helmken, H., Wheaton, W., Baity, W.A., and Peterson, L.E.: 1978, Nature 272, 679.
Maraschi, L., and Treves, A.: 1977a, Astrophys. J. 218, L113.
Maraschi, L., and Treves, A.: 1977b, Astron. Astrophys. 61, L11.
Morrison, P.: 1958, Nuovo Cimento 7, 858.
Paul, J.A., Bennett, K., Bignami, G.F., Buccheri, R., Caraveo, P., Hermsen, W., Kanbach, G., Mayer-Hasselwander, H.A., Scarci, L., Swanenburg, B.N., and Wills, R.D.: 1978, Astron. Astrophys. 63,L31.
Ruffini, R.: 1972, *Black Holes*, p. 451, Eds. C. de Witt and B.S. de Witt, Gordon and Breach, New York.
Scarsi, L., Bennett, K., Bignami, G.F., Boella, G., Buccheri, R., Hermsen, W., Koch, L., Mayer-Hasselwander, H.A., Paul, J.A., Pfeffermann, E., Stiglitz, R., Swanenburg, B.N., Taylor, B.G., and Wills, R.D.: 1977, Proc. 12th ESLAB Symp., ed. Battrick, B., ESTEC Noordwijk, ESA SP 124, 3.
Strong, A.W.: 1977, Nature 269, 394.
Swanenburg, B.N.; 1978, submitted to Astron. Astrophys.
Swanenburg, B.N., Bennett, K., Bignami, G.F., Caraveo, P., Hermsen, W., Kanbach, G., Masnou, J.L., Mayer-Hasselwander, H.A., Paul, J.A., Sacco, B., Scarsi, L., and Wills, R.D.: 1978, Nature, in press.
Swanenburg, B.N.: 1977, Proceedings of the Int. School of General Relativistic Effects, MPI-PAE/Astro 138, 361.
Taylor, J.H., and Manchester, R.N.: 1975, Astron. J. 80, 794.
Taylor, J.H., and Manchester, R.N.: 1977, Astrophys. J. 215, 885.

COMPACT RADIO SOURCES (SUMMARY)

K.I. Kellermann
Max-Planck-Institut für Radioastronomie, Bonn, F.R.G.

Compact radio sources are found in the nuclei of spiral and elliptical galaxies, radio galaxies, and quasars. Generally they show a complex distribution of radio brightness with two or more spacially separated components, which is remarkably similar to that of the giant extended sources, but which is smaller by five to six orders of magnitude (e.g. Kellermann 1978).

In general the compact radio sources are variable with outbursts in flux density on time scales of a year or so and with corresponding changes in the brightness distribution. In some sources, such as the quasars 4C39.25 and the radio galaxy NGC 1275 the components remain stationary, but in others the components appear to separate with apparent velocities of up to 10 times the speed of light (e.g. Cohen et al. 1977, Shaffer et al. 1977). Although there are many ad hoc attempts to interpret this "superluminal" motion, there are as yet no completely satisfactory explanations.

REFERENCES

Cohen, M.H. et al.: 1977, Nature 268, 405.
Kellermann, K.I.: 1978, Physica Scripta 17, 257.
Shaffer, D.B. et al.: 1977, Astrophys. J. 218, 353.

Bengt E. Westerlund (ed.), Stars and Star Systems, 15.

CHEMICAL EVOLUTION OF GALAXIES

B.E.J. Pagel
Royal Greenwich Observatory, Herstmonceux, Sussex, U.K.

ABSTRACT

The chemical evolution of disk galaxies is discussed with special reference to results obtained from studies of the oxygen abundance in H II regions. Normal spirals (including our own) display the by now well known radial abundance gradient, which is discussed on the basis of the simple enrichment model and other models. The Magellanic Clouds, on the other hand, and the barred spiral NGC 1365, have been found to have little or no abundance gradient, implying a very different sort of evolution that may involve large-scale mixing. Finally, the simple model is tested against a number of results in H II regions where the ratio of total mass to mass of residual gas can be estimated. It turns out to fit adequately the Magellanic Clouds and a number of H II regions in the outer parts of spiral galaxies, but in more inner parts it fails, as do more sophisticated models involving infall during the formation of galactic disks that have proved very successful in other respects.

1. INTRODUCTION

The chemical evolution of galaxies is part of the wider problem of galactic evolution in general and as such cannot be reviewed with any sort of completeness in a single lecture. I have therefore selected for special treatment just one very limited aspect of the subject, namely some recent data on oxygen abundances in H II regions which reveal the presence of large-scale radial abundance gradients in many disk-like galaxies, but their absence in some others, and I shall try to discuss the interpretation of these effects in term of the simple enrichment model, mentioning at the same time one or two of the more sophisticated models which are superior to the simple model in the context of the general problem of galactic chemical evolution, but which offer essentially the same interpretation of abundances in H II regions as it does.

Bengt E. Westerlund (ed.), Stars and Star Systems, 17–31.

To place the discussion in its context, however, I begin with some more general remarks about the construction of models to account for the evolution of heavy-element enrichment in galaxies.

The basic postulates of such models are, first that galaxies are formed somehow by the coming together of intergalactic gas on a shorter or longer time-scale accompanied by star formation, again on a shorter or longer time-scale, and secondly that the gas (and the stars being born from it at any particular time) is initially lacking or at least poor in heavy elements from carbon upwards, and is gradually enriched in such elements by material coming from stars that have completed their own evolution and eject the products of nucleosynthesis in the course of a violent or lingering death. In the case of elliptical galaxies and the bulge and halo population of at least the earlier spirals, the spheroidal shape suggests that star formation occurred on a time-scale that was shorter, or at least no longer, than that of the collapse of the system as a whole, and was largely completed a long time ago; whereas in disk-like systems star formation has evidently been delayed for some reason, such as perhaps low density or tidal effects from the central nucleus, so that substantial amounts of gas are still there and we can see new stars being born at the present time.

Many different sorts of observations are relevant to the problem of the enrichment of the interstellar medium (ISM) in heavy elements. These include

1. The cosmic abundance distribution and its quasi-universality (cf. Unsöld 1974), that is the fact that in stars that we believe to be unevolved, and therefore to show in their spectra the chemical composition with which they were born, the relative abundances of carbon and heavier elements are not very different, even when the total heavy-element abundance relative to hydrogen varies by large factors. There are, in fact, well-established exceptions to this rule (cf. Pagel 1978), notably nitrogen and to a lesser extent oxygen and the elements formed by slow neutron captures (*s*-process), but it is certainly quite a good approximation to regard the total heavy-element abundance Z as being proportional to the metallicity revealed by spectroscopy and photometry of stars.

2. The age dependence of stellar metallicities in the solar neighbourhood, which is still rather controversial apart from the well-known metal deficiency of the globular clusters and the ultra-high-velocity stars of the galactic halo population which are certainly old (Eggen, Lynden-Bell and Sandage 1962). However, a positive correlation between metallicity and time of formation of disk stars probably also exists (Mayor 1976).

3. The statistical distribution of metallicities in nearby stars, also known as Schmidt's G dwarf problem, which has proved such a stumbling block to simple models of galactic enrichment (van den Bergh 1962; Schmidt 1963; Pagel and Patchett 1975; Lynden-Bell 1975; Audouze and

Tinsley 1976).

4. The large-scale radial abundance gradient in our Galaxy, deduced from the metallicity and kinematics of nearby stars (Mayor 1976; Janes 1977) and from oxygen abundances in H II regions (Peimbert 1978 and references therein).

5. Similar large-scale abundance gradients found in other large galaxies, both elliptical and spiral, from the integrated colours and spectra of stellar populations (Faber 1977 and references therein) and from observations of H II regions (Searle 1971 and discussion below). Most large galaxies share the property of having greater metallicity or heavy-element abundance in their central regions than in the outer parts.

6. Abundance differences between giant and dwarf elliptical galaxies, in the sense that the metallicity in the central regions increases steadily with the mass or luminosity (Faber 1973, 1977). This effect, and the large-scale gradient in elliptical galaxies, probably result from a systematic flow of enriched gas from newly-formed stars, both inwards towards the centre and outwards in the form of galactic winds, during the formative stages (Larson 1974).

7. Isotopic abundances in the Solar System and in the interstellar medium, notably the galactic centre (Vigroux *et al.* 1976).

8. The abundances of the light elements Li, Be, B (Reeves and Mayer 1978).

9. Small, but significant variations in the ratio of helium to hydrogen (Peimbert and Torres-Peimbert 1976) which presumably reflect a fairly mild "topping up" by stellar activity of the bulk of the helium, which was formed in the "Big Bang".

2. INGREDIENTS FOR CHEMICAL EVOLUTION MODELS

What, then, are the basic ingredients of models for the chemical evolution of galaxies?

1. First there are initial conditions; mostly one assumes that we begin with gas left over from the Big Bang containing helium 4 and 3, deuterium, and perhaps lithium 7, but some models appeal to a "prompt initial enrichment", possibly due to a previous generation of massive stars either in the disk or in the halo, which leads to an initially significant abundance of heavy elements.

2. Next one needs a set of models of the end-products of stellar evolution. Which stars synthesise and eject what elements after how much time and in what quantities? It is generally believed that the most massive stars eject so-called primary nucleosynthesis products

which can be built up directly from hydrogen and helium in the course of stellar evolution without the intervention of extensive mixing processes, e.g. C, O, Ne and most of the nuclear species up to the iron group. Certain other elements like N and *s*-process elements could also be primary if the necessary mixing processes occur, or they could be secondary in the sense that they can only be produced by stars of later generations that were already born with C, O and metals present, which are then changed into N and neutron sources, with metals (especially iron) acting as seed nuclei for the *s*-process, in the course of hydrogen and helium burning.

It is also generally agreed that, within the galactic lifetime of the order of 10^{10} years, stars of intermediate mass between one and a few solar masses eject their excess of material above the white dwarf residue as gas that may have undergone some secondary processing, but no primary processing, and therefore return gas to the interstellar medium without enriching it in oxygen or metals. Finally, stars of about one solar mass or less simply form and remain there without returning anything significant to the ISM, as do the white dwarfs and other compact remnants such as neutron stars left behind by more massive stars after their death.

3. The third ingredient is the initial mass function (IMF), originally introduced by Salpeter (1955), which describes the relative rates at which the stars of different masses are born. Talbot and Arnett (TA 1973a) have put together a combination of IMF and models of end-products of stellar evolution that is by no means definitive, but is a sort of "Identikit" model that is useful, and has been widely used in galactic chemical evolution calculations because it provides all the data that are needed in a convenient form. The two most important parameters of the combined stellar evolution-IMF model are firstly the fraction of mass in each generation of stars that is returned to the ISM, which I shall call β, and secondly the "yield" of heavy elements, which I shall call p, defined as the total mass fraction of primary synthesis products ejected in each generation, relative to the fraction $1-\beta$ that remains locked up in small, long-lived stars or collapsed remnants. In the "Instantaneous Recycling" approximation (TA 1971; Searle and Sargent 1972), where one assumes that all of the relevant stellar evolutionary processes take place instantaneously compared to the time-scale of galactic evolution, these two numbers are constants characteristic of the IMF adopted; e.g. in the TA Identikit model, $1-\beta \simeq 0.8$ and $p \simeq 0.01$, but they will vary, either in more accurate models taking into account the errors in instantaneous recycling, or if the IMF is assumed to vary in some systematic way.

4. The fourth ingredient is a model of star formation as a function of time, gas mass, gas density and probably other parameters. This involves a vast amount of interesting and unknown physics (cf. Larsson 1977), the problem being to account for the fact that we see both ellipsoidal systems with stars and usually little gas, in which it

seems that star formation has apparently been very efficient, and flat disc-like systems having up to about half the mass still in the form of gas in which either star formation is relatively inefficient or new gas is being continually supplied. In these systems, new star formation appears to accompany the advent of some kind of large-scale shock, e.g. material passing through a spiral shock in our own Galaxy (Roberts 1969; Oort 1974) or tidal interaction between galaxies (Toomre 1974) or a collision between a galaxy and an intergalactic gas cloud (Fosbury and Hawarden 1977), but how to quantify these effects and extrapolate them back into the past is quite a problem, and many authors have simply assumed that the rate of star formation is proportional to a small simple power of either the density or the total mass of gas.

One advantage of the instantaneous recycling approximation is that for many purposes, like the discussion of abundance gradients or abundance differences between galaxies, or the statistical distribution of metal abundances in stars seen today, the laws of star formation are not critical to the results, because they are subsumed in the present ratio of stars to gas which can be directly measured.

5. The fifth and final ingredient contains the assumptions made as to all the other relevant processes in galactic evolution besides the birth and death of stars. Two questions are important in particular:

a. Does chemical enrichment take place in an isolated zone, or do we have to take into account the interaction with the environment in the form of inflow or outflow of material?

b. Is our zone well mixed, or does it have a random scatter in chemical composition at any time, or are there systematic effects like Metal Enhanced Star Formation (MESF: TA 1973b; Talbot 1974; TA 1975) that cause stars to differ in their average metallicity from the general interstellar medium?

These questions bring us back again to fundamental problems in galaxy formation and star formation respectively, and they can be answered in the long run only in the context of comprehensive theories of these two processes; but the study of the distribution of abundances can give us a few clues as to things that may have happened.

3. THE SIMPLE MODEL AND ITS SUCCESSES AND FAILURES: ABUNDANCE GRADIENTS

Much attention has been devoted to the simplest model of galactic chemical evolution and its failure, notably in connection with the notorious problem of the narrow distribution of metallic abundances in the G dwarfs of the solar neighbourhood. In this context, and in many others, it is clear that the simple model fails, but there are enough contexts where it does not obviously fail that - just because it is simple - it is of interest to look at them.

What is the simple model? This is the model that assumes star formation to take place in an isolated and well-mixed zone, initially consisting of pure gas with no heavy elements, according to an invariable initial mass function. The zone in question could be either a whole galaxy - if this is believed to be well mixed - or a cylindrical coaxial shell perpendicular to the plane of a flattened galaxy - if each zone can be considered to evolve in isolation from neighbouring zones. Then in the instantaneous recycling approximation, the yield of primary elements is p = const. and their abundance (by mass fraction) in the gas or in newly formed stars at any particular time is given by

$$Z = p \ln\left(1 + \frac{s}{g}\right) \qquad (1)$$

where s is the mass locked up in stars (including compact remnants) and g the mass of gas that is left (including gas that has been through stars and ejected one or more times) (Searle and Sargent 1972). Searle and Sargent pointed out that this equation predicts a large-scale abundance gradient in galactic disks, provided that we assume evolution to take place separately in concentric cylindrical zones, because $\frac{s}{g}$ (and hence Z) generally decreases as we go outwards from the central regions. The effect can readily be quantified (cf. Shields and Searle 1978) if we assume that the total surface density, (s + g), decreases outwards exponentially (Freeman 1970):

$$s + g = \frac{\alpha^2 M}{2\pi} e^{-\alpha R} \qquad (2)$$

where M is the total mass of the disc, R is the radial distance from the centre and α is a constant; and that the surface density of gas is more or less constant, which is a fair approximation over quite a wide range of R to the H I distribution in the Magellanic Clouds and the Scd galaxies M33 and M101 (Pagel *et al.* 1978) and probably even to the distribution of total hydrogen including molecules in our own Galaxy (Cesarsky *et al.* 1977; Serrano 1978). - A very different hydrogen distribution was derived from the survey of galactic CO emission by Gordon and Burton (1976), but this is probably affected by an abundance gradient in carbon relative to hydrogen. - In this case we have

$$Z = \text{const.} - p\alpha R \qquad (3)$$

(where the const. = $p \ln\frac{\alpha^2 M}{2\pi g}$) and the abundance decreases linearly with the radial distance.

Equation (3) can be tested for the solar neighbourhood using data for galactic H II regions assembled by Peimbert (1978) on the basis of observations by himself and others. (Data for other galaxies will be discussed later.)

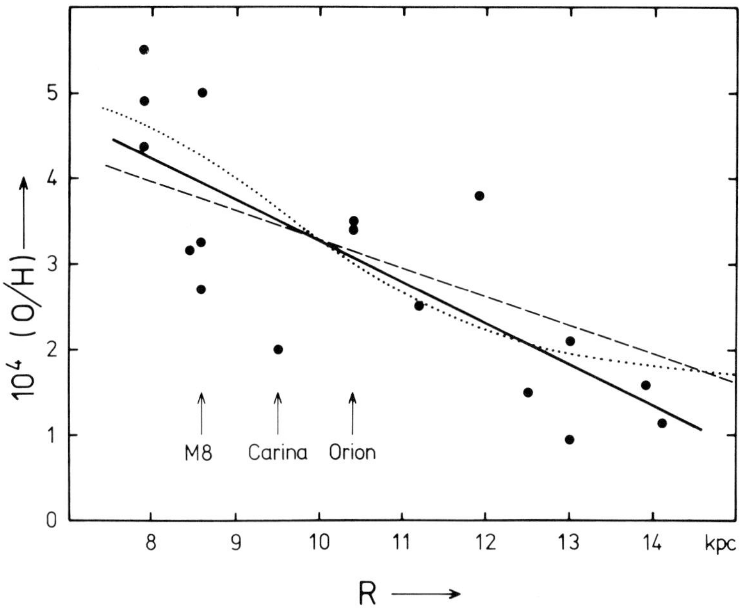

Fig. 1 Oxygen abundances in galactic H II regions after Peimbert (1978), but adjusted to zero temperature fluctuation, plotted against galactocentric distance R (assuming $R_\odot = 10$ kpc). Full line: least squares solution (10^4 (O/H) = 8.06 - 0.48 R). Broken line: Larson (1976) Model 6 Variant (Tinsley and Larson 1978). Dotted curve: MESF model (Talbot and Arnett 1975).

The result is shown in Fig. 1, where it is seen that, within the considerable scatter of the observations, the trend is indeed linear between 8 and 14 kpc from the Galactic centre (assuming $R_\odot = 10$ kpc), over a factor of 3 in oxygen abundance. A very similar trend in stellar CN abundances has been found by Janes (1977). The solid line shows a simple least-squares solution and from the slope of the graph (-0.48×10^{-4} kpc^{-1}) and the exponential disc radius of Schmidt's (1965) model, $\alpha = 0.28$ kpc^{-1} (Serrano 1978), we can use equation (3) to derive the yield (assuming Z = 25 (O/H))

$$p = 0.004$$

which is a little less than the canonical value of 0.01 in the Identikit model, but agrees within the uncertainties (cf. Tinsley 1976). (My oxygen abundances and slope are lower than found by Peimbert because I have assumed zero temperature fluctuations within the H II regions.)

Of course, the fact that the simple model fits these data does not mean that it is correct. In fact we know it to be incorrect in the solar neighbourhood because of the G dwarf problem and other models have been developed that are vastly superior to the simple model both because they fit the G dwarfs better and because they form part of a more comprehensive physical picture. The predictions of two such models, scaled to cross the least-squares line at 10 kpc, are shown in Fig. 1. The broken line represents a model of slow formation of galactic disks by gradual infall of unprocessed material due to Larson (1976; Tinsley and Larson 1978) referred to by the authors as Model 6 Variant, in which infall is cut off after a while, whereas the dotted curve represents a metalenhanced star formation (MESF) model by TA (1975). The first one differs from the simple model by not assuming evolution in isolated zones, while the second one differs in assuming that stars are born systematically richer in heavy elements than the average of the interstellar medium. These more sophisticated models are also fairly successful in accounting for the abundance gradient in this region, but the essential physics that accounts for this particular effect is already implied in the simple model; in fact its success in this particular application imposes some constraints on the more elaboratate models.

What is the underlying physics behind the abundance gradient?

The fundamental effect is that, as we look across the Galaxy at this particular moment of time, more of the gas has been changed into stars in the inner regions than in the outer regions, and this in turn has received two types of explanation. Some authors have considered rates of star formation varying with a power law of the average gas density, with an exponent greater than 1, which can arise from a variety of reasons like the free-fall time scale, the rate of collisions of clouds and so forth. Such laws, when applied to the past history of the solar neighbourhood using the simple model, come into conflict with the attempt to reconcile the relative numbers of large-mass and low-mass stars seen today with a constant and smooth IMF: there are too few long-lived dwarf stars, compared with the number of short-lived O and B stars, to permit the average past rate of star formation to have exceeded the present rate by as much as would be required by a power law in the gas surface density with an exponent of even 1, let alone more than 1 (Tinsley 1976; Mayor and Martinet 1977; Serrano 1978). There are several ways out of this dilemma: the IMF could have varied (there is positive evidence that in some cases it does, e.g. Freeman 1977); it could be discontinuous, with low-mass and high-mass stars being born in quite separate sets of events (Eggen 1976; Smith *et al.* 1978); and, more simply, the mass of gas could have had a phase in the past when it was increasing owing to infall (Larson 1974; Lynden-Bell 1975; Larson 1976) so that one can have a power law in the gas density for the rate of star formation combined with a non-monotonic dependence on time. As Lynden-Bell has shown very elegantly, this form of departure from the simple model provides a natural explanation for the narrow range of abundances in G dwarfs; Larson has pointed out, further, that it can also account for the indications that the past rate of star formation has been fairly uniform, so that Larson's (1976) model with decaying infall, and the closely related analytical model of Lynden-Bell, seem to provide the "best buy" in current models for the evolution of disc-like galaxies.

However, there is another important effect first pointed out by Oort (1974). Star formation is associated with the shock to which interstellar material is subjected when it passes through a spiral density wave (Roberts 1969) and so the rate of star formation should depend on the difference $\Omega(R) - \Omega_p$ between the local angular velocity of rotation and the constant pattern speed of the spiral wave. Current data on the rate of formation of massive stars in different parts of our Galaxy give some support to a dependence on $\Omega - \Omega_p$ as well as the total surface density of gas (Serrano 1978). Jensen, Strom and Strom (1976) have related abundance gradients in a number of late-type spirals to $\Omega - \Omega_p$ and other calculated parameters of the spiral shock; but this result has to be viewed with some caution, because the abundances we see today depend on the integrated effect of all of past history, during which (as we have just seen) the density of gas at different places in any one galaxy, and perhaps even the rate of passage through spiral shocks, must have varied in quite a complicated way.

4. GALAXIES WITH LITTLE OR NO ABUNDANCE GRADIENT

We have seen that abundance gradients in disk-like galaxies can give useful information about their chemical (and other) evolution. It was this consideration that motivated a systematic survey of H II regions in the Magellanic Clouds, particularly the Small Cloud, by Pagel, Edmunds, Fosbury and Webster (1978) using the Anglo-Australian Telescope with the Robinson-Wampler Image Dissector Scanner and the Boksenberg Image Photon Counting System, both of which are multichannel instruments that enable one to observe extensive regions of the spectrum in a short time, and also making use of the abundances found in MC H II regions by other authors.

There are several reasons why the Magellanic Clouds are of special interest. They resemble late-type spirals in having an exponential-disc-like surface brightness distribution and the Large Cloud has a well-defined rotation curve. The Small Cloud is very rich in gas and has strong non-circular motions. Finally, both systems are probably losing material in the Magellanic Stream (Mathewson *et al.* 1974), if this results from tidal interaction with our own Galaxy (Fujimoto and Sofue 1976; Lynden-Bell 1976; Davies and Wright 1977), and if this material is preferentially gas, then it could have the effect of reducing the effective yield.

The upshot of our investigations is given in Figure 4 of Pagel *et al.* (1978). In the Large Cloud we find a small and ill-determined gradient in the oxygen abundance of

$$-(0.17 \pm 0.12) \times 10^{-4} \text{ kpc}^{-1},$$

a factor of 3 less than in the solar neighbourhood, although the exponential disk length α^{-1} is only 1.6 kpc (Freeman 1970), about half that of the Galaxy in our own neighbourhood. In the Small Cloud, the result is unambiguous: there is no gradient whatsoever. The Small Cloud is evidently a well-mixed system, which is not surprising in view of the non-circular motions and its generally chaotic character, while the Large Cloud may retain some individuality in concentric cylindrical zones, but probably not very much.

What, then, happens in galaxies of other morphological types? So far, using the AAT in collaboration with D.E. Blackwell, M.S. Chun, M.G. Edmunds and G. Smith, we have studied some H II regions in the Southern Scd galaxy NGC 300, which displays a similar sort of abundance gradient to those previously observed by Smith (1975) and by Shields and Searle (1978) in M33 and M101. But what is probably a more interesting result comes out from our study of H II regions in another system, the well-known barred spiral NGC 1365, having a "hot spot" nuclear region (Sersic and Pastoriza 1965) in which non-circular motions were observed by Burbidge, Burbidge and Prendergast (1962). The spectrum of the nucleus was studied by Osmer, Smith and Weedman (1974) who found it to be like that of a conventional H II region, heavily reddened by internal dust, but with quite normal abundances. However, it also seems to contain a (variable) X-ray source (Ward *et al.* 1978). We have observed three H II regions at different locations at one end of the bar and along a spiral arm and the significant result is that all of these H II regions have close to solar abundances of both oxygen and nitrogen, with no large-scale gradient present (Fig. 2). One barred spiral, like one swallow, does not make a summer, especially when our swallow is such an active one. However, the fact that there are also bar-like structures in the Magellanic Clouds is suggestive, and one may venture a conjecture that in barred galaxies abundance gradients are more or less smoothed out by non-circular motions (cf. Edmunds 1977). If so, then the distribution of abundances can indeed give us further useful indications on the structure and evolution of galaxies, but there is evidently a lot more work to be done.

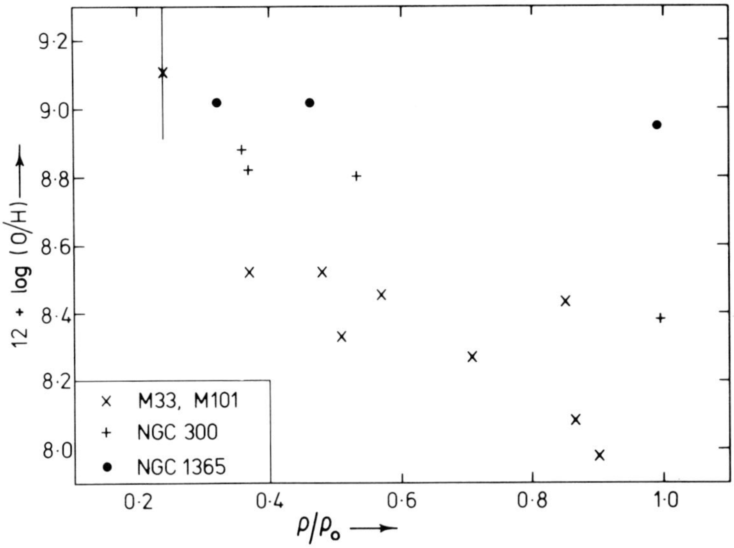

Fig. 2 Logarithm of oxygen abundance as a function of galactocentric distance (in units of the 25^m $arcsec^{-2}$ isophotal radius) for M33 and M101 (Smith 1975, adjusted to zero temperature fluctuation; Shields and Searle 1978), and for the Scd spiral NGC 300 (similar to M33) and the SBbc galacy NGC 1365 (Blackwell *et al.* 1978).

5. THE FINAL CRUNCH

From the point of view of the simple model, the absence of an abundance gradient in systems like the Magellanic Clouds is not particularly embarrassing, because we can simply define our unit of isolated evolution to be the whole galaxy rather than a concentric cylindrical zone. So we now ask ourselves: if we do this, can we fit the oxygen abundance in the H II regions of the Magellanic Clouds, using equation (1), to one another and to the abundances in the H II regions of our own Galaxy and of M33 and M101 for which data on the surface densities of gas and total mass are available from optical and radio observations? Figure 3 shows the result of an attempt to do this, in which I have updated and modified Fig. 5 of Pagel *et al.* (1978) by incorporating the new data of Shields and Searle (1978) for M101 and those of Peimbert (1978) for our own Galaxy, and using for the latter Schmidt's mass model and the total hydrogen distribution found by Cesarsky *et al.* (1977), scaled up by a factor of 2.5 to give a surface density of 10 $m_\odot$ pc^{-2} at the solar distance (Pagel and Patchett

1975; Serrano 1978).

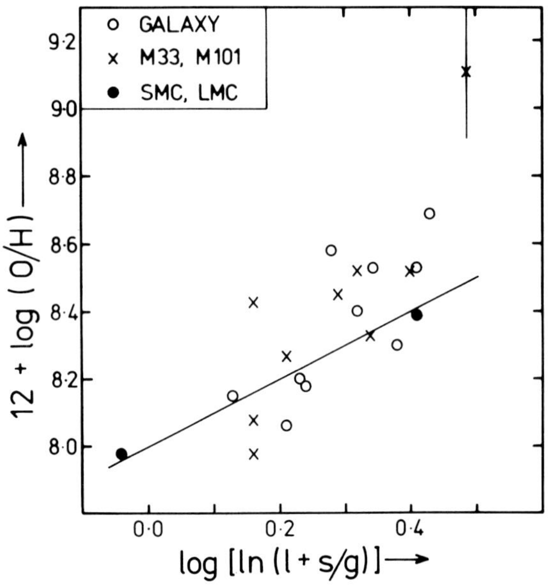

Fig. 3 Logarithm of oxygen abundance as a function of the "astration" parameter log [ln $(1 + \frac{s}{g})$] corresponding to equation (1). The relation predicted by the simple model is a straight line of unit slope. The line in the diagram corresponds to a yield of about 0.003.

The straight line in Fig. 3 represents the predictions of the simple model, now using a logarithmic scale. We can see that it fits both Magellanic Clouds and some of the H II regions in the outer parts of our own Galaxy and M33 and M101 (the ones on the left side of the diagram). This result is of some interest, because it suggests that the progress of element formation in the Magellanic Clouds has not been very different from what it has been in the outer parts of the three spiral galaxies, so that modifications to the simple model, with inflows, outflows or changes in the IMF have had only minor effects. The abundances in these H II regions, in very different sorts of galaxies, seem to be fixed by the ratio of residual mass of gas to total mass more or less in accordance with equation (1), and there are no significant differences between the Magellanic Clouds and the corresponding zones of the classical spiral galaxies.

Going over to the right side of the diagram, however, in regions where $\frac{s}{g} > 10$ or so, the abundances increase much faster than predicted, and here we reach the crunch where the simple model breaks down; nor does it seem that the more sophisticated models that I have mentioned are any better, because the gradients that they predict are about the same or even flatter than in the simple model. Likewise errors in the instantaneous recycling approximation, which begin to be significant here, can only make matters worse, because the accurate calculations always lead to somewhat lower abundances than instantaneous recycling predicts, because of unprocessed matter coming out of stars of intermediate mass as time goes on. Therefore, as Shields and Searle have rightly pointed out, quite a radical departure from the simple model seems to be needed, and the kind of departure needed is a more drastic one than in the otherwise successful models such as those of Larson that I have referred to. What the departure should be is not clear; one possibility is prompt initial enrichment of just the inner regions by supernova ejecta from the halo as proposed by Ostriker and Thuan (1975), but there are, of course, others. In any case, despite the comparative crudeness from which galactic enrichment models still suffer, I am convinced that with the aid of further observations and theoretical studies the abundances in H II regions will tell us a great deal more about the general problem of the evolution of galaxies in the next few years.

I am grateful to D.E. Blackwell, M.S. Chun, M.G. Edmunds, and G. Smith for their part in the studies of NGC 300 and 1365 reported here; to M. Peimbert for communicating his results in advance of publication; to A. Serrano, whose D. Phil. dissertation has helped to clear up my ideas on a number of topics; and to the SRC Panel for the Allocation of Telescope Time for allocating observing time on the AAT and to the Director and staff of the Anglo-Australian Observatory for their cooperation.

REFERENCES

Audouze, J., and Tinsley, B.M.: 1976, Ann. Rev. Astron. Astrophys. 14, 43.
van den Bergh, S.: 1962, Astron. J. 67, 486.
Blackwell, D.E., Chun, M.S., Edmunds, M.G., Pagel, B.E.J., and Smith,G: 1978, in preparation.
Burbidge, E.M., Burbidge, G.R., and Prendergast, K.H.: 1962, Astrophys.J. 136, 119.
Cesarsky, C.J., Casse, M., and Paul, J.A.: 1977, Astron. Astrophys. 60, 139.
Davies, R.D., and Wright, A.E.: 1977, Mon. Not. R. astron. Soc. 180, 71.
Edmunds, M.G.: 1977, I.A.U. Coll. no 45, p. 67.
Eggen, O.J.: 1976, Q.J.R. astron. Soc. 17, 472.
Eggen, O.J., Lynden-Bell, D., and Sandage, A.R.: 1962, Astrophys. J. 136, 748.
Faber. S.M.: 1973, Astrophys. J. 179, 731.
: 1977, in *The Evolution of Galaxies and Stellar Populations*, Eds. B.M. Tinsley and R.B. Larson, Yale University, p. 157.
Fosbury, R.A.E., and Hawarden, T.G.: 1977, Mon. Not. R. astron. Soc. 178, 473.
Freeman, K.C.: 1970, Astrophys. J. 160, 811.
: 1977, in *The Evolution of Galaxies and Stellar Populations*, p. 133.
Fujimoto, M., and Sofue, Y.: 1976, Astron. Astrophys. 47, 263.
Gordon, M.A., and Burton, W.B.: 1976, Astrophys. J. 208, 346.
Janes, K.A.: 1977, I.A.U. Coll. no 45, p. 173.
Jensen. E.B., Strom, K.M., and Strom, S.E.: 1976, Astrophys. J. 209, 748.
Larson, R.B.: 1974, Mon. Not. R. astron. Soc. 166, 585.
: 1976, Mon. Not. R. astron. Soc. 176, 31.
: 1977, in *The Evolution of Galaxies and Stellar Populations*, p. 97.
Lynden-Bell, D.: 1975, Vistas in Astron. 19, 299.
1976, Mon. Not. R. astron. Soc. 174, 695.
Mathewson, D.S., Cleary, M.N., and Murray, J.D.: 1974, Astrophys. J. 190, 291.
Mayor, M.: 1976, Astron. Astrophys. 48, 301.
Mayor, M., and Martinet, L.: 1977, Astron. Astrophys. 55, 221.
Oort, J.H.: 1974, IAU Symp. no 58, p. 375.
Osmer, P.S., Smith, M.G., and Weedman, D.W.: 1974, Astrophys. J. 192, 279.
Ostriker. J.B., and Thuan, T.X.: 1975, Astrophys. J. 202, 353.
Pagel, B.E.J.: 1978, *Liège Astrophysical Colloquium: Les Eléments et leurs Isotopes dans l'Univers*, in press.
Pagel, B.E.J., Edmunds, M.G., Fosbury, R.A.E., and Webster, B.L.: 1978, Mon. Not. R. astron. Soc., 184, 559.
Pagel, B.E.J., and Patchett, B.E.: 1975, Mon. Not. R. astron. Soc. 172, 13.
Peimbert, M.: 1978, I.A.U. Symp. no 84, in press,
Peimbert, M., and Torres-Peimbert, S.: 1976, Astrophys. J. 203, 581.

Reeves, H., and Mayer, J.-P.: 1978, Preprint.
Roberts, W.W.: 1969, Astrophys. J. 158, 123.
Salpeter, E.E.: 1955, Astrophys. J. 121, 161.
Schmidt, M.: 1963, Astrophys. J. 137, 758.
: 1965, in *Galactic Structure*, Eds. A. Blaauw and M. Schmidt, Univ. of Chicago Press, p. 513.
Searle, L.: 1971, Astrophys. J. 168, 327.
Searle, L., and Sargent, W.L.W.: 1972, Astrophys. J. 173, 25.
Searle. L., and Shields, G.A.: 1978, Astrophys. J. 222, 821.
Serrano, A.: 1978, Thesis, University of Sussex.
Sersic, J.L., and Pastoriza, M.: 1965, Publ. astron. Soc. Pacific 77, 287.
Smith, H.E.: 1975, Astrophys. J. 199, 591.
Smith, L.F., Biermann, P., and Mezger, P.: 1978, Astron. Astrophys. 66, 65.
Talbot. R.J.: 1974, Astrophys. J. 189, 289.
Talbot, R.J., and Arnett, W.D.: 1971, Astrophys. J. 170, 409.
: 1973a, Astrophys. J. 186, 51.
: 1973b, Astrophys. J. 186, 69.
: 1975, Astrophys. J. 197, 551.
Tinsley, B.M.: 1976, Astrophys. J. 208, 797.
Tinsley, B.M., and Larson, R.B.: 1978, Astrophys. J. 221, 554.
Toomre, A.: 1974, I.A.U. Symp. no 58, p. 347.
Unsöld, A.: 1974, Proc. First Europ. Astron. Meeting, Athens 1972, vol. 3 *Galaxies and Relativistic Astrophysics*, Eds, B. Barbanis and J.D. Hadjidémetriou, Springer-Verlag, Berlin.
Vigroux, L., Audouze, J., and Lequeux, J.: 1976, Astron. Astrophys. 52,1.
Ward, M.J., Wilson, A.S., Penston, M.V., Elvis, M., Maccacaro, T., and Tritton. K.P.: 1978, Astrophys. J. 223, 788.

HIGH-ENERGY PHENOMENA ON THE SOLAR SURFACE

Claudio Chiuderi
Istituto di Astronomia, Università di Firenze, Italy

1. INTRODUCTION

I shall start this review of the high-energy phenomena on the solar surface by briefly discussing the meaning of the title. By solar surface it is meant the whole of the solar atmosphere, from, say, $\tau_{5000} = 1$ up. By high-energy phenomena I mean the complex of the observed events that involve the presence of suprathermal particles and the emission of radiation whose brightness temperature exceeds the kinetic temperature. High-energy phenomena take place in a sporadic way and generally follow the cycle of solar activity. The most outstanding manifestations are called the solar flares. I emphasize again that with this word I intend to encompass all the consequences of a sudden release of energy in a well defined portion of the solar plasma, such as the enhanced emission of radiation over the whole electromagnetic spectrum and the acceleration of particles up to the GeV range. Since, generally speaking, radiation of different wavelengths originate from different levels of the solar atmosphere, the conspicuousness of the flare phenomenon over the entire electromagnetic spectrum implies that a large vertical portion of the atmosphere itself is involved.

On the scale of total energy output from the sun flares are absolutely irrelevant. In fact the total energy emitted by a large flare is estimated of the order of a few units in 10^{32}, with a total duration of about one hour and a total surface area involved of a few units in 10^{-4} of the solar surface. Keeping in mind the figures of $L_\odot \simeq 4 \times 10^{33}$ erg s^{-1} and $F_\odot \simeq 6.3 \times 10^{10}$ erg cm^{-2}s^{-1} for the sun's total luminosity and flux, we see that

$L_{flare} \simeq 10^{-5} L_\odot$ and $F_{flare} \simeq (10^{-1} - 10^{-2}) F_\odot$, thus proving the

above statement. Given their overall irrelevance to the global energetics of the sun, why then should we study the flares? Apart from the obvious answer that physicists and astronomers just study everything, I believe that flares are a good example of a class of

Bengt E. Westerlund (ed.), Stars and Star Systems, 33–50.

phenomena widespread in astrophysics: the violent conversion of some form of energy into heat and kinetic energy of accelerated particles. The proximity of the sun makes it possible to study this process in much better detail in the case of flares than, say, in the case of radiosources. I do not know if the information gained through a proper understanding of the flare phenomenon could then be directly scaled to other astrophysical contests, but it appears to me that this is certainly a sensible first step towards a *quantitative* description of high-energy phenomena in astrophysics.

As everybody knows, the literature on flares is very vast and the rate of papers on this subject has dramatically increased after the Skylab Mission. There is no hope, therefore, to cover even a modest part of it in a short talk. I must consequently make a selection, that, like all selections, will be inevitably biased. I shall start by giving an extremely short and incomplete description of the observed characteristics of flares, simply to provide the basic facts that have to be explained by physical theories. I shall then concentrate on the two problems that are the core of the physical understanding of the flare phenomenon: the build-up of energy prior to the flare and the mechanism of primary energy release. A third vital problem, that of the particle acceleration, will be only mentioned in passing. I will make no attempt to give a complete set of references. There are available a number of recent extensive review articles and books (Tandberg-Hanssen 1967; Öhman *et al.* 1968; Newkirk 1974; Sakurai 1974; Kane 1975; Priest 1976; Svestka 1976 a,b; Rust 1978) to which the reader is referred for a complete reference list.

OBSERVATIONS OF FLARES

As already mentioned the total energy release in a flare is insignificant compared with the total solar luminosity, which is mostly concentrated in the optical band. This explains the rarity of flares seen in white light. The first flare ever seen, however, was a white-light flare, observed by Carrington on September 1, 1859. Observations show that the continuum of the solar spectrum is hardly affected by flares, while the most prominent absorption lines show strong variations, thus proving the chromospheric rather than photospheric nature of the optical flare. The most detailed optical studies of flares have been made in the H_α line. The temporal behaviour of the intensity shows a rapid rise, followed by a slower decay. The rising period is called the flash phase, the declining one, the main phase. As can be seen from a comparison of the time development of a flare at different wavelengths, the flash phase appears to be the most interesting part of the flare since all the energetic processes are taking place during this phase. The typical duration of the flash phase is of the order of 100 seconds. The main phase may last ten or hundred times more. Emission in H_α is usually concentrated in patches, that often have the appearance of two parallel strips (two-ribbon flare). Flare emission, morfologically similar to that in H_α, is present in most of the other chromospheric lines from about 1500 Å to about 1 μ.

The feasibility of observations from space has immensely widened our capability of probing the nature of the flare phenomenon, by making accessible large portions of the electromagnetic spectrum,

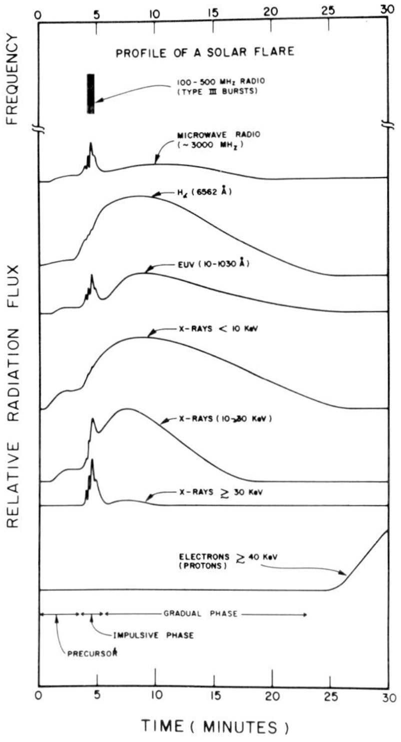

Figure 1. Typical time development of the electromagnetic and particle emission of a flare.

which are absorbed by the terrestrial atmosphere. The shift of interest from chromospheric flares, once the only observable ones, to coronal flares reflects our new capabilities. Observations in the EUV range have been performed with high resolution during the Skylab Mission. Emission lines between 50 and 1500 Å show dramatic enhancements during flares in coincidence with the flash phase. From these observations we learned that the emission is confined to a series of loop-like structures, with the hottest material ($T \simeq 1 - 2 \times 10^7$ K) at the top of the loop. The loops are filled with cooler coronal material ($T \simeq 2 - 5 \times 10^6$ K) linking two low-temperature ($T \simeq 1 - 5 \times 10^4$ K) chromospheric ribbons. The emission of soft X-rays (0.1 - 10 keV) during a flare is also enormously increased, exceeding sometimes that of the quiet corona by a factor of 10^4. The morphology in soft X-rays is similar to that in EUV, but the time development is more reminiscent of the H_α flare. The emission of hard X-rays is concentrated in a very short time interval (impulsive phase).

At the other extreme of the electromagnetic spectrum we have flare-associated radio emission from microwave to metric wavelengths. Particularly interesting are the so-called type-II bursts which are interpreted as radioemission from a plasma excited by the passage through the corona of a shock wave at a velocity of thousands of km/sec. Associated with flares are also the ejections of large amounts of mass. The shocks and the ejecta can carry with them up to three quarters of a flare's total energy, the remaining part going into visible, EUV and X-ray radiation.

Of particular importance is the association of flares with the structure of the solar magnetic field. Flares occur only in regions of strong magnetic fields, called active regions. Active regions are usually, but not exclusively, associated with sunspots. The photospheric value of the magnetic field is of the order of thousands of gauss. The field decreases with height to a rather uncertain coronal value between 10 and 100 G. Using the visible structures of the chromosphere and lower corona as tracers of the field lines and combining this with the knowledge of the value of the line-of-sight photospheric field, $B_{||}$, deduced from magnetograms, we come to a picture of the large-scale magnetic field distribution. The main building blocks of this structure are loop-like configurations, that intersect and group together, forming systems of magnetic arcades, that join regions of opposite polarity. Flares invariably start close to the so-called neutral line, that is the line where $B_{||} = 0$. They also start preferably in regions where the magnetic field shows a complicated, tangled structure. After the flare the field shows sometimes a simpler pattern, but no major changes are normally observed at photospheric level. Particles are accelerated during flares. Their presence is witnessed by the emission of microwave and hard X-ray bursts and is directly detected at the Earth's orbit through the observed modification of the state of the ionosphere or the increase of the cosmic ray flux. Acceleration of electrons and ions to 100 keV is a standard feature of flares, but the so called "proton flares" produce protons up to 100 MeV and the "cosmic-ray flares" may boost heavy particles in the GeV energy range. By interacting with the ambient nuclei in the solar atmosphere, these high-energy particles can initiate nuclear reactions. In this case, production of gamma rays is likely to ensue via the radiative decay of excited nuclei, or positron annihilation, or deuterium formation. Observations have been successfully performed of the 2.23 MeV line from $p + n \rightarrow d + \gamma$, of the 0.51 line from $e^+ + e^- \rightarrow \gamma + \gamma$ and of the 4.43 MeV and 6.14 MeV de-excitation lines of C^{12} and O^{16}.

From the wealth of observations that I have just quickly described, it is possible to extract a reduced set of morphological characteristics and temporal behaviours that define what I will call the "standard flare". Let me briefly enumerate them. The flare is preceded by a passive phase in which energy accumulates. This phase has a variable duration but can be as short as one day (from the interval between omologous flares, i.e. flares that occur at the same location with similar morphological and temporal behaviours). The flare onset is signaled by bursts of impulsive radio and hard X-ray emission (same pre-heating of the chromosphere prior to the bursts is also observed). Chromospheric H_α emission, EUV radiation from the chromosphere-corona transition zone and coronal soft X-ray emission rise rapidly in a typical timescale of 100 seconds. H_α and soft X-rays then enter in a slowly decaying phase. Considerable amounts of plasma are ejected in space and shock waves form accompanied by radio emission. Particles are accelerated to high energies. Flares are ignited in active regions close to the location where the line-of-sight component of the photospheric magnetic field vanishes. The fundamental coronal structures which participate to the flare are loops or arcades, presumably delineating the magnetic field lines.

Even in this oversimplified description the tantalizing variety of phenomena that comprise a flare clearly emerges. Let us now review the attempts that have been made to interpret them.

FLARE PHYSICS

From our (arbitrary) definition of a "standard flare" it appears that the physical system we are dealing with is a magnetized plasma in which a large amount of energy is suddenly released in an explosive fashion. The state of the plasma is certainly a non-equilibrium one and is generally described as a turbulent state. To understand the physics of such a system is a very difficult task. Before proceeding to the discussion of specific problems it is important to realize that most of the progress, which can be expected in the future in this field, will come from a massive, quantitative application of plasma physics to the flare problem. Of course, all flare theories are based on plasma physics, but only recently an increasing number of astrophysicists became aware of the close relationship of the laboratory studies of plasmas in fusion-oriented devices to solar plasmas. The recent spectacular advance in laboratory plasma physics has produced methods and results, whose importance can hardly be overestimated, which just wait to be applied, intelligently, not blindly, to solar physics and more generally to astrophysics.

To limit ourselves to very general problems, it is clear that we will have to answer at least such basic questions as:

i) in which form is the flare energy stored in the pre-flare state?

ii) which is the actual mechanism that allows a relatively slow, passive accumulation of large amounts of energy in a metastable state?

iii) what triggers the flare?

iv) which mechanisms operate during the flash and main phases to convert the stored energy into kinetic energy of accelerated particles, heat and radiation?

The first two questions are generally referred to as the energy build-up problem, the last two as the energy release problem. Both problems have received a great deal of attention: in spite of this it is only fair to say that our understanding of the flare mechanism is still far from complete. However, even if the answers to all the questions listed have not been given, it has been possible to identify a few firm physical points that will certainly be the basis of the theory of solar flares which is presently being formulated.

Energy build-up

Turning to the examination of the first of the fundamental problems listed above, it is now generally accepted that the energy that will eventually fuel the flare is stored in the form of magnetic energy. Alternative sources, typically thermal or gravitational energy, appear to be unable to satisfy the energy requirements of flares.

Let us consider the amount of thermal energy that can be stored in a flaring region. Since the photosphere is only marginally affected by the flare, we must conclude that the energy has to be stored at small optical depths, say $\tau < 0.01$, where the pressure scale height is substantially smaller than the thickness of the flaring region. In this situation, the thermal energy is stored essentially in a region occupying the same surface area as the flare but being only one scale height thick. Using for the pressure and scale height at $\tau = 0.01$ the values given by the semi-empirical models of the solar atmosphere (such as HSRA) we get for typical flare sizes an energy of about $3 - 5 \times 10^{30}$ ergs. The gravitational energy is of the same order of magnitude, so we conclude that these forms of energy give at most 0.1 of the required energy, even neglecting consideration of the efficiency of the conversion process.

On the contrary, magnetic energy easily fulfils the energy requirements: a field of 160 G, occupying a volume of about 10^{29} cm^3, already gives 10^{32} ergs. Of course, one must specify something more on magnetic fields, since we must be able to utilize the energy contained in the field to convert it into heat and kinetic energy of accelerated particles. For instance, if the energy is contained in a potential ($\vec{j} = 0$) field, there is no way of extracting this energy if we keep the boundary conditions fixed (which seems dictated by the observations of irrelevant photospheric changes connected with flares). In fact, the potential field is the one that possesses the minimum energy among all the fields that satisfy the same boundary (photospheric) conditions. We therefore need some "free energy" above that stored in potential fields. In identifying these non-potential fields we are helped by the knowledge that the chromospheric and coronal structures in the flaring region appear to be rather stable prior to the flare and certainly so on the MHD timescale, of the order of a few seconds in coronal conditions. Therefore these structures appear to be MHD stable and MHD equilibrium must hold:

$$0 = \vec{\nabla} p + \frac{1}{c} (\vec{j} \times \vec{B}) = \vec{\nabla} p + \frac{1}{4\pi} (\vec{\nabla} \times \vec{B}) \times \vec{B}. \qquad (1)$$

If we assume equal scales for the pressure and field variations, the ratio of the first-to-second term in the above equation is, in order of magnitude:

$$(p/L)/(B^2/4\pi L) = \frac{1}{2} p/(B^2/8\pi) = \frac{1}{2}\beta ,$$

where β is the usual plasma physics parameter. Since in the high solar atmosphere $\beta \ll 1$, we are entitled to consider force-free fields:

$$\vec{j} \times \vec{B} = \frac{4\pi}{c} (\vec{\nabla} \times \vec{B}) \times \vec{B} = 0$$

or

$$\vec{\nabla} \times \vec{B} = \alpha \vec{B} . \qquad (2)$$

It is the presence of field-aligned currents that increases the energy content of the force-free fields above that of the current-free potential field, and it is precisely this energy that is available for the flare.

There have been successful attempts to check in a quantitative way this idea. On August 1972, three major flares developed in a large active region. This region was observed continuously in the H_α line and magnetograms of the underlying photospheric field were taken. The location of the flares was on both sides of the magnetic neutral line between two intense spots. Bright structures in the form of arches were seen, connecting regions of opposite magnetic polarity. Assuming that these arches delineate the magnetic field lines, they may be compared with computer-generated field lines, obtained by reconstructing a 3D force-free or potential field from the measured vertical component of the photospheric field given by the magnetograms. Tanaka and Nakagawa (1973) used a single term Fourier representation of the line-of-sight field B_z:

$$B_z = B_o \sin(\pi x/L_x) \cos(\pi y/L_y) \; ; \quad (-\pi/2 \le x,y \le \pi/2) .$$

The x-axis connects the centers of opposite polarity and the parameters B_o, L_x and L_y are estimated from the observations. The field equations (2), with constant α can then be solved and the magnetic energy computed. The result is:

$$W = \frac{B^2}{16\pi} \frac{L_x L_y}{k \cos\delta}$$

where

$$k^2 = (\pi/L_x)^2 + (\pi/L_y)^2 ,$$

and δ is the (observable) angle that the projected field makes with the x-axis at the origin. The relationship between α and δ is:

$$\alpha = k \sin \delta ,$$

showing that for a potential field ($\alpha = 0$) the energy is minimum and that the energy content increases with the shear measured by δ. Tanaka and Nakagawa found that the amount of extractable energy for the August 7 flare was 6×10^{32} ergs and that the shear was gradually increasing in the pre-flare phase, whereas δ showed a rapid decrease ($\simeq 35^o$) after the flare. Similar analysis of other flares produced consistently the same general picture.

Therefore, we shall accept the notion that the energy for flares is stored in the form of stressed, almost force-free, magnetic fields and that the field relaxes to a quasi-potential form when the extra energy has been used up. The ultimate energy source for flares resides, of course, in those photospheric or sub-photospheric motions that are responsible for the generation of currents in the higher levels of the solar atmosphere. We can envisage the whole process of energy accumulation in the pre-flare state as the gradual production of currents along the field lines due to the motions of the feet of the lines themselves, anchored in the highly conducting, high-β photospheric plasma. The problem of dynamical evolution of force-free fields has been attacked in a series of papers by Nakagawa and his co-workers (for a short review see Nakagawa 1978, and references therein). The problem presents serious mathematical difficulties, since the actual form that the field assumes depends not only on the surface distributions of currents and fields but also on the specific dynamical history traversed by the system to reach the final state. Depending on the particular way in which the foot points are moved, the field takes the form of low and flat arches or of more extended and elongated structures. Interestingly enough, all the motion patterns examined seem, however, to be capable of providing the canonical value of 10^{32} ergs in the appropriate period of time.

The non-uniqueness of the solution of the force-free equation has prompted a number of very interesting investigations on the stability of sheared force-free fields. In fact, from the existence of multi-valued solutions, the possibility arises of discontinuous topological changes of the field structure, and the association of these abrupt modifications with the eruptive onset of a flare is clearly a very attractive possibility. It can be shown that the solution of the problem of the dynamical evolution of a two-dimensional force-free field in the ideal MHD scheme is equivalent to the solution of

$$\nabla^2 A = -\lambda F(A)$$

where the field is determined by A, B = B (A), F(A) is a given, arbitrary function of A and λ is a parameter related to the displacement of the foot-points. Choosing a potential field as the initial state, λ can be taken as zero at the start, increasing values of λ measuring the departure from the potential condition. By using a particular form of F(A), Birn *et al.* (1978) have shown that two topologically distinct solutions can be found from $\lambda = 0$ to a certain value $\lambda = \lambda^*$. Only one of these solutions, however, has physical relevance since the other contains field lines disconnected from the photospheric boundary. The physical solution was shown to be stable by these authors while no solution was found beyond $\lambda = \lambda^*$. These facts together suggest the identification of $\lambda = \lambda^*$ with the threshold for the shearing motion of the foot-points of field lines, beyond which a discontinuous change takes place that might be interpreted as the flare onset. More recent work (Heyvaerts, private communication) seem to indicate the existence of a third solution for all values of λ, corresponding to an open field topology. Surpassing the threshold $\lambda = \lambda^*$ could then mean a discontinuous ("catastrophic") transition from a closed field topology to an open one. The direct application of these results to the flare problem is probably still premature, considering the oversimplified description used (two-dimensional geometry, no dissipative effects). The interest of these type of investigations is, however, undoubtable and comparison with the observed changes in the magnetic field structure due to the occurence of a flare should be made.

Energy release

We turn now to the problem of how the stored magnetic energy is released, or, in other words, how is it possible to convert magnetic energy in other forms of energy, notably heat and kinetic energy of particles. If the magnetic energy has to decrease, part of the magnetic field has to disappear and this is very difficult to achieve in a very conducting plasma like the solar one. Actually the extremely high value of the electrical conductivity of the solar atmosphere has been one of the major stumbling blocks on the way of building a physical theory of the solar flares.

If the conductivity is generated through coulomb collisions between electrons and ions, we can use the so-called Spitzer conductivity (Krall and Trivelpiece 1973)

$$\sigma = 3.2 \times 10^6 \ T^{3/2} \ (\text{e.s.u.})$$

which for relevant solar temperatures can reach the value of 10^{16} e.s.u., comparable to the conductivity of the best metallic conductors at room temperature. The magnetic induction equation is:

$$\frac{\partial B}{\partial t} = \vec{\nabla} \times (\vec{v} \times \vec{B}) + \eta \nabla^2 B \tag{3}$$

where

$$\eta = \frac{c^2}{4\pi\sigma} .$$

This equation clearly displays two characteristic timescales, depending on the ratio of the two terms on the right hand side, which defines the magnetic Reynolds number:

$$R_m = (Lv)/\eta .$$

At large values of R_m the first term dominates; we have the phenomenon of the frozen-in fields and the characteristic timescale is the Alfvén transit time

$$\tau_A = L/v = L/c_A ,$$

since in this situation the typical velocity is simply the Alfvén velocity

$$c_A = B/(4\pi\rho)^{\frac{1}{2}} .$$

For relevant solar values this time turns out to be of the order of one second or less. To the other extreme of small R_m we have the resistive decay of the magnetic field, due to the ohmic dissipation of the sustaining currents. The characteristic timescale here is:

$$\tau_D = L^2/\eta ,$$

whose value for solar conditions can be as large as 10^{10} seconds, or about 300 years. It is clear that neither of these two timescales is suitable for the description of a flare, whose characteristic time lies in between. It is also clear on the other hand, that the resistive dissipation of the magnetic field energy offers the most direct and viable way for conversion to different energy forms. Therefore, in order to speed up the dissipative processes we are forced to consider situations in which the typical spatial scale, ℓ, is much smaller than the flare region size, L, or in which the electrical conductivity is substantially reduced below the classical case, or both.

Considering the first possibility let us assume that the resistive decay of the field actually takes place in a region of size $\ell \ll L$. Such a region, where the current density $j = (c/4\pi)(B/\ell)$ is much larger than elsewhere and the scale for energy conversion much shorter, is called a current sheet. Crossing the current sheet the magnetic field rotates and may even reverse its sign. A possible scenario for our process of magnetic energy conversion is therefore the following. The field initially occupies a region of linear dimension L. The plasma flows toward the resistive region with velocity v. Due to the high conductivity the field is effectively frozen into the plasma and it is therefore carried along with it. If the energy conversion is assumed to be a stationery process, in the resistive region R_m must be of the order unity, which physically means that the amount of field that is being dissipated must exactly balance the amount of field convected by the plasma into the resistive region. The typical timescale for the release of the magnetic energy in this steady situation is therefore

$$\tau = L/v \; ,$$

which we identify with the flaring time, $\tau_f = 10^3$ sec. Since on the other hand in the resistive region $R_m = 1$, by eliminating v, we have

$$1 \cong \ell v/\eta \cong (L\ell)/(\eta\tau_f)$$

or

$$\ell \cong \eta \, \tau_f/L \; .$$

If we use Spitzer's conductivity and $L = 10^9$ cm, we get for ℓ values well below one centimeter! We are therefore forced to conclude that the resistivity, η, must be much higher than its classical coulomb value. The ways in which this could be achieved will be considered later on.

The difficulty just outlined by this order-of-magnitude calculation remains there even if more refined models of energy dissipation are considered. A well-known example of these is the Petschek model, proposed in 1964. The model starts with a two-dimensional magnetic field, with an X-type neutral point, that is also a stagnation point of the flow. The neutral point is the center of a small region where a balance between the diffusion time and the external hydromagnetic (Alfvén) time is obtained by an extreme reduction of the width of the central region. It is assumed that four slow MHD shocks originate from this region. These shocks would normally propagate upstream with the oblique Alfvén velocity, but they are kept in a fixed position by the incoming flow. The shock fronts are the locations where actually most of the energy deposition occurs, whereas the central reconnecting region just provides the correct field topology that allows the formation of the shock system. The Petschek model has gone through a number of criticisms and improvements both on physical and mathematical aspects. Even if considerable success has been obtained, especially by Priest and his co-workers (see Priest 1976, for a review and reference list) in solving the difficult mathematical problems connected with this model, the reconnection rate remains so far too low to cope with the flare requirements. The main objection against this model is that it does not explain how the shock system is formed but just assumes it from the start and checks that such an assumption is consistent with the equations. Being a stationary model, with considerable free energy, its stability against MHD (or resistive) perturbations is a particularly important and unanswered problem. For, if it could be shown that the whole configuration is unstable over relatively short timescales, serious doubts could be raised about the very existence of the stationary Petschek configuration. Directly related to the above problem is the fact that the Petschek mechanism is not a canonical instability in the plasma physics sense, namely something that grows in time. This reflects a basic choice in the attempt to solve the complicated set of nonlinear space and time dependent equations of resistive MHD. To reduce to an acceptable level the difficulties of the solution one has two basic possibilities: either to ignore the time derivatives and to restrict himself to stationary situations (as done in the Petschek case), or to linearize the equations and investigate the time development of the small perturbations. This second approach leads to mechanisms based on the existence of resistive instabilities, to which we now turn our attention.

As already mentioned, the decoupling of the magnetic field from the plasma appears to be the necessary condition to allow the utilization of the free energy stored in the stressed field. We have already seen, however, that the simple resistive decay is a far too slow process to be of interest for flares. Resistive magnetic tearing, which we are going to describe, offers a very interesting possibility of speeding up the process of energy conversion. To be effective, the magnetic tearing instability requires a sheared magnetic field without need of a neutral point or layer. To simplify the discussion we will consider the case of plane geometry, as done in the pioneering work of Furth *et al.* (1963). Let us assume an initial sheared field,

$$\vec{B}_o = B_{ox}(y)\,\vec{e}_x + B_{oz}(y)\,\vec{e}_z \quad . \tag{4}$$

In each y = const. plane, the field lines are straight, but their inclination changes with y. Consider then perturbations of the form

$$f_1(r,t) = f_1(y)\exp(ikx + \omega t) \tag{5}$$

for all first order quantities. For simplicity we have chosen the x-axis along the k-vector of the perturbation. Linearizing the Maxwell equations:

$$\vec{\nabla} \times \vec{E} = -\frac{1}{c}\frac{\partial B}{\partial t} \quad ,$$

$$\vec{\nabla} \times \vec{B} = \frac{4\pi}{c}\,\vec{j} \quad ,$$

Ohm's law,

$$\eta\vec{j} = \vec{E} + \frac{1}{c}\vec{v} \times \vec{B} ,$$

and the momentum equation,

$$\rho \frac{\partial\vec{v}}{\partial t} = \frac{1}{c}\vec{j} \times \vec{B} ,$$

we easily get (v and E are first order quantities)

$$\rho\omega v_y = \frac{1}{c\eta}(E_z B_{ox} - \frac{1}{c} v_y B^2_{ox}) \quad . \tag{6}$$

Assume now detachment between fluid and field, i.e. that the fluid moves but not the field lines, so that $\vec{E} = 0$. Equation (6) then proves the existence of a restoring force that becomes infinitely large as $\eta \to 0$, which is just another way to demonstrate the frozen-in situation of infinite conductivity. This restoring force, however, is arbitrarily weak near the points where B_{ox} vanishes. The first term in Equation (6) can be shown to be a destabilizing one, and, due to its weaker dependence on B_{ox}, is the dominant one near $B_{ox} = 0$. For unrestricted choices of the reference system, the above result can be restated by saying that we may expect that detached fluid motion can take place in a small region about a null point of $\vec{k}\cdot\vec{B}_o$. Outside that region the second term dominates and the field, which is effectively frozen-in, is convected towards the null layer by the fluid motion. In the resistive region, whose dimension is entirely self-determined in this problem by the appropriate matching of the boundary conditions, the plasma is broken loose from the magnetic field, which then tears and reconnects, forming elongated magnetic cells. This configuration has a lower magnetic energy and a higher thermal and kinetic energy than the initial one. Working out the mathematical details, we find for the growth time, $\tau \sim 1/\omega$,

$$\tau \sim k^{2/5}\tau_A^{2/5}\ \tau_D^{3/5} \ .$$

This shows that this is typically a long-wavelength instability, whose linear growth time is intermediate between the Alfvén transit time and the resistive decay time, a very important feature for application to solar flares. While the rigorous treatment of the tearing mode instability can easily become prohibitively involved if situations more realistic than that so far considered are investigated, the basic physics is simple, since it reflects the tendency of a uniform current sheet to be unstable against filamentation. As a result of this instability the final state will be that of a series of separate current filaments. But in this case, the magnetic field lines must circulate around the current filaments and this implies a change from the initial field topology, that can be only accomplished if the freezing condition is violated somewhere.

Having given a very sketchy description of the basic physics behind the tearing mode instability, let me briefly review the degree of success of its application to the flare problem. The *linear* growth time can be barely made adequate for flaring time requirements, using Spitzer's conductivity. However, one should not compare this time with observations, but rather investigate the nonlinear time development of this instability and compare the nonlinear saturation time with the observed timescales. This has been done only for special field configurations in plane geometry (Van Hoven and Cross 1973), with comforting results. In fact, the instability is seen to saturate in about five linear timescales, which shows that at least in this case the needed increase of resistivity above its classical value is not very large. On the other hand, from studies of laboratory plasmas, it is known that the resistive instabilities growth rates are strongly geometry-dependent, so that calculations in realistic cases are badly needed. The width of the resistive layer, although small ($\simeq$ 10 m, in coronal conditions), is certainly much more reasonable than the widths encountered in Petschek-type models. The electric field generated by the instability appears to be large enough to provide the primary acceleration of the high-energy particles.

The total amount of the initially stored energy that is converted into other forms, has also been computed by Van Hoven and Cross, who found that up to 30 per cent of the energy is removed. The same authors also suggest that the nonlinearly saturated stage of the tearing mode instability may mimic some of the features of the stationary Petschek model, thus providing a link between these two main classes of models.

We have seen that in any case a resistivity larger than the classical coulomb one is likely to be required. This can be achieved if the plasma becomes turbulent, since in this case drastic alterations of the transport coefficient take place. The Spitzer conductivity in particular is replaced by the so-called turbulent conductivity. The new physics here is brought about by the fact that the direct electron-ion collisions, responsible for the classical conductivity, become less efficient than those between the electrons and plasma excitations (quasi-particles). In the few cases in which details have been worked out (Langmuir turbulence and ion-sound turbulence), the conductivity is lowered by substantial factors. A reduction of five orders of magnitude is typical. There are, however, problems in this case, too. In fact, let us ask how the turbulence forms. It is generally thought that the turbulent plasma, namely a plasma in which a particular type of excitation is present in excess of that expected in a thermal plasma, is a consequence of the fact that the current density surpasses some critical value. In plasma-physics jargon it is said that we are in the presence of current-driven instabilities which produce the particular excitation we are considering. However, by virtue of the Maxwell equations, a high current means a high field gradient, and since the field cannot be assumed arbitrarily high, high gradients imply small scale lengths. By computing the numbers, one realizes that the problem of very small spatial scales can be again a very severe one. An example of this is given by a recent calculation by Heyvaerts and Kuperus (1978), who study the triggering of plasma turbulence during the compression of a current sheet. Such a process is likely to occur when new magnetic flux emerges from the photosphere into the corona and presses against the pre-existing field. These authors show that the conditions for the production of microturbulence can be satisfied some time after the beginning of the compression, when the sheet shrinks to something of the order of one centimeter.

To conclude this section on the primary energy release problems it is important to recall that the process of magnetic field reconnection, which in my opinion stands presently as the best candidate for the mechanism of energy conversion needed to fuel a flare, is being given a large amount of attention also in laboratory plasma studies. It is suspected, in fact, that these types of processes are responsible for the destructive disruptions which occur inside the devices of the controlled- fusion research. Chances are that under the combined attacks of plasma physicists and astrophysicists some substantial progress might be in sight.

FLARE MODELS

We now leave aside the discussion of the basic physical processes which pertain to the flare energetics and discuss a few more phenomenological models that have been proposed in recent years.

The first of these models is especially due to Lin and Hudson (1976) and tries to develop a consistent picture of the energy release during a flare, by postulating that everything happens as a consequence of an accelerated beam of electrons that impinges on the chromosphere. In a sense this model complements our discussion in the preceding sections, since no assumption is made concerning the primary energy release process that accelerates the electrons in the first place, but the consequences of the interaction of the accelerated electrons are closely examined. As already mentioned, the emission of hard X-rays is normally interpreted as bremsstrahlung emission from non-thermal electrons. From the observed X-ray spectrum and the assumption that the electrons lose all their energy in the flare region ("thick target model"), it is possible to compute their energy distribution, the total number of electrons and the total energy involved. Typical values for large flares are 10^{38} electrons and 10^{31} - 10^{32} ergs. These figures show that the energy content of the electron beam is well in the range of the total energy output of a large flare.

Let us ask now how the solar atmosphere reacts to the sudden deposition of a large amount of energy. The electron beam which is supposedly directed downward, explosively heats the various layers encountered. From the temporal sequence of the flare emission at different wavelengths, it seems likely that the bottom of the transition zone, $T \lesssim 6 \times 10^4$ K, is the first to react to the impulsive electron heating. The correctness of this idea has not been checked so far, due to the lack of simultaneous, high-resolution observations in EUV and hard X-rays. It would be tempting to associate the impulsive emission with the first bright patches of Hα emission, which then would indicate the location of impact of the electrons on the chromosphere. Again, resolution problems make it difficult to check this idea. Due to the peculiar form of the radiative energy loss function of an ionized plasma, we have two different behaviours of the lower and upper solar atmosphere, the discriminating temperature being ≈ 6×10^4 K. Below this temperature it is possible to radiate away the energy deposited by the electron beams, so that these layers can remain in an equi-librium state, whereas at higher temperatures the radiative efficiency decreases, so that it is no longer possible to dispose of the extra energy by radiation. The atmosphere then undergoes a rapid expansion, whose visible consequence would be the ejection of large clouds of material. These in turn could generate shock waves in the corona, giving rise to type-II radiobursts. Mechanisms have been proposed to accelerate protons in the shock fronts generated by large flares up to the GeV range (Sturrock 1974).

The main objections to the thick target model are related to the requirements that the model poses on the acceleration mechanism that supposedly creates the beam of accelerated electrons and by the stability of the beam itself. In fact, to produce 10^{38} electrons which hit the chromosphere over a limited area in a period $\lesssim$ 100 seconds, requires an extremely efficient acceleration mechanism and a rather dense electron population. There are no known coronal structures that appear to be capable to satisfy these requirements. Moreover, if we just have a stream of electrons rushed down from the corona, the magnetic field generated by this current would be so strong as to destabilize the beam in a time much shorter than the observed duration of the impulsive phase. This field could be counter-balanced by a return current opposite to the beam. Here, however, we will probably run into difficulties with the two-stream instability. Attractive as it may be, the thick target model seems to need at least substantial modification to overcome the difficulties just mentioned.

Another type of model, the emerging flux model (Heyvaerts *et al.* 1977) starts from the observation of the strong correlation between the emergence of new photospheric flux and the occurrence of flares. The observed rate of flux emergence seems to be sufficient to produce at least the smaller flares, through reconnection processes with the pre-existing magnetic field. During the compression of the newly emerged field against the old one, long current sheets form that can stimulate energy conversion, via Petschek-type mechanisms. This energy is certainly sufficient to produce the observed pre-flare heating and this is a positive feature of the model. Investigations along this line had already been done by Syrovatskii (1969) and

Coppi (1975). Heyvaerts *et al.* follow the dynamical evolution of the sheet and show that the critical threshold for the onset of turbulence is likely to be surpassed at a certain stage. From their calculation, however, it appears that the energy release rate is too small to account for large flares. An interesting, so far unexplored, possibility, would be to incorporate resistive magnetic tearing in the flux emergence model, to help increasing the energy conversion rate.

As a final model, I will briefly discuss the so-called "unstable arch model", especially advocated by Spicer (1976). This starts from the consideration that from the Skylab pictures we know that most, if not all, coronal flares take place in loop structures. The similarity of these structures with the diffuse cylindrical or toroidal pinches of fusion-oriented machines (especially with the so-called "reversed z-pinch"), suggests to apply well-known methods of laboratory plasma physics to the coronal structures. In a sense the situation is ideal, since the magnetic field is sheared in these cylindrical or semi-toroidal geometries and offers quite naturally the possibility of increasing the magnetic field gradient beyond a critical threshold. This can be done simply by increasing the twist of the field lines, controlled in turn by the motion of the foot-points. Tearing mode instabilities may then set in, releasing the energy contained in the stressed field lines. Since the critical layers where the instability develops are again those where $\vec{k} \cdot \vec{B}_o = 0$, it is clear that if a perturbation has non-vanishing components over a certain range of k-values, the reconnection could take place in several different locations, thus greatly increasing the overall energy release. It is also possible to envisage an interaction among the various unstable layers, that again results in an increased energy production. The unstable arch model offers a number of interesting possibilities, most of which, however, have just been stated as such, with no quantitative work associated. So far, the only calculations done have been those referring to the MHD stability of the coronal loops (Chiuderi *et al.* 1977, Giachetti *et al.* 1977), a necessary preliminary to the study of the more delicate resistive instabilities.

As a final remark, I would like to stress once again that the construction of a reliable flare model or theory is a complicated, little intuitive and very challenging problem. Simplifications are necessary, even over-simplifications sometimes. But qualitative or hand-waving arguments should never be considered a substitute for a serious calculation.

REFERENCES

Birn, I., Goldstein, M. and Schindler, K.: 1978, Solar Phys. 57, 81.
Chiuderi, C., Giachetti, R. and Van Hoven, G.: 1977, Solar Phys.54, 107.
Coppi, B.: 1975, Astrophys. J. 201, 735.
Furth, H.P., Killeen, J. and Rosenbluth, M.N.: 1973, Phys. Fluids 6, 459.
Giachetti, R., Van Hoven, G. and Chiuderi, C.: 1977, Solar Phys. 55, 371.
Heyvaerts, J., Priest, E.R. and Rust, D.M.: 1977, Astrophys. J. 216, 123.
Heyvaerts, J. and Kuperus, M.: 1978, Astron. Astrophys. 64, 219.
Kane, S.R.: 1975, *Solar Gamma-, X and EUV Radiation*, Reidel Publ. Co., Dordrecht, Holland.
Krall, N.A. and Trivelpiece, A.W.: 1973, *Principles of Plasma Physics*, McGraw-Hill, New York.
Lin, R.P. and Hudson, H.S.: 1976, Solar Phys. 50, 153.
Nakagawa, Y.: 1978, *Pre-flare Magnetic fields, one-free fields and evolution*, ATM Flare Workshop Report.
Newkirk, G., ed.: 1974, *Coronal Disturbances*, Reidel Publ. Co., Dordrecht, Holland.
Ohman, Y., Hosinsky, G. and Kusoffsky, U.: 1968, *Mass Motions in Solar Flares and Related Phenomena*, Interscience-Wiley, New York.
Petschek, H.E.: 1963, *AAS-NASA Symposium on the Physics of Solar Flares*, p. 425, NASA SP-50, U.S. Government Print Office, Wash., D.C.
Priest, E.R.: 1976, Solar Phys. 47, 41.
Rust, D.M.: 1978, *Solar Flares* (to be published in *Solar System Plasma Physics*, Eds. C.F. Kennel, L.J. Lanzerotti, and E.N. Parker.
Sakurai, K. : 1974, Astrophys. Space Sci. 28, 375.
Spicer, D.S.: 1976, NRL Report 8036, Naval Research Laboratory, Wash., D.C.
Sturrock, P.A.: 1974, *Coronal Disturbances*, p. 437, Ed. G. Newkirk Jr, Reidel, Dordrecht, Holland.
Svestka, Z.: 1976, Solar Phys. 47, n. 1.
Svestka, Z.: 1976, *Solar Flares*, D. Reidel, Dordrecht, Holland.
Syrovatskii, S.I.: 1969, *Solar Flares and Spaces Research*, p. 346, Eds. C. De Jager and Z. Svestka, North-Holland, Amsterdam.
Tanaka, K. and Nakagawa, Y.: 1973, Solar Phys. 33, 187.
Tandberg-Hanssen, E.: 1967, *Solar Activity*, Blaisdell Waltham, Mass, USA.
Van Hoven, G. and Cross, M.: 1973, Phys. Rev. A7, 1347.

OBSERVATIONAL AND THEORETICAL ASPECTS OF GAS IN EARLY TYPE GALAXIES

C.A. Norman
Laboratory Astrophysics, Huygens Laboratorium, University of Leiden, The Netherlands

ABSTRACT

Observational searches for neutral hydrogen in early type galaxies and their environments are reviewed and the theoretical motivation for such searches is discussed. The 9 positive detections of HI in ellipticals are considered, as well as processes which might explain the absence of gas from ellipticals in general. Accretion of intergalactic clouds, the wind accretion interaction and related effects such as star formation and activity in the nucleus are analysed. In assessing the origin of S0's as due to gas removal from normal spirals, implications of relatively recent ($z \lesssim 0.5$) gas removal from galaxies are proposed for both galaxy evolution and the density evolution of radio sources.

I. INTRODUCTION

There is very little gas in elliptical galaxies. The upper limits of surveys place the mass of HI in most ellipticals at less than $\lesssim 10^8\ M_\odot$ (Knapp *et al.* 1978a); the ionised component is negligible ($\lesssim 10^6\ M_\odot$). The interstellar medium in such galaxies therefore comprises only less than approximately 0.1% of the total mass.

Actual detection of HI in ellipticals is extremely rare, there being only 9 bona fide instances to date. Ionised gas is more frequently detected; but this is still in only 15% of ellipticals (Osterbrock 1960).

The low limits of observed gas content for ellipticals are in contradiction to current theoretical estimates of mass loss from evolved stars in this class, assuming that the gas remains in the galaxy in the form of HI (Faber and Gallagher 1976). Transformation of the gas into unobservable forms, such as molecular hydrogen or very hot gas ($> 10^6$ K), seems unlikely. Thus, a number of mechanisms

Bengt E. Westerlund (ed.), Stars and Star Systems, 51–66.

have been proposed to explain the removal of gas from ellipticals: consumption of gas by star formation; expulsion of gas by stellar winds; evaporation of gas driven by thermal conduction effects; and sweeping of gas out of a galaxy if the external medium exerts a significant ram pressure on the gas internal to the galaxy (Section III). The S0's provide a particularly important test for some of these mechanisms, if one believes that S0's are spirals which have had their gas removed (Gisler 1978).

The observable presence of HI in a few ellipticals, therefore, necessitates special consideration. It appears that either the removal mechanisms have been less efficient in these cases than in the great majority of ellipticals; or else that the gas content is the result of accretion processes. The best evidence for accretion comes from the particular cases of NGC 4278 (Gallagher *et al.* 1977, Bottinelli and Gouguenheim 1977a, Knapp *et al.* 1978a) and the Spindle Galaxy NGC 2685 (Shane 1978). Primordial hydrogen clouds in groups, mass transfer in binary systems, and close encounters which strip gas from galaxies are possible sources for the accreted gas (Section IV).

Detection of HI in the 9 ellipticals is important for two reasons. Firstly, a rotation curve has already been obtained for the HI distribution in NGC 4278 and thus an estimate made of the mass-to-light ratio outside the optical galaxy. This estimate provides evidence for a massive non-luminous halo. It is expected that the same procedure can be applied to the other gas-containing ellipticals. Secondly, this small group of ellipticals appears to give a correlation between non-thermal radio source activity in the nucleus and the HI content of ellipticals (Ekers 1978). There is a real possibility that we may now be observing accretion phenomena in nearby ellipticals that can give considerable guidance to the understanding of the physics of the energy sources for active nuclei (Section V).

Before discussing these specific objects (II B) and theoretical problems (III-V), let us consider the results of recent surveys.

II. OBSERVATIONS

A. Surveys

The classic work of Minkowski and Osterbrock (1959), Osterbrock (1960), and Morgan and Osterbrock (1969), demonstrated that about 15% of ellipticals and 30 - 50 percent of S0's show {OII} λ 3727, which is probably associated with the nucleus. The typical mass of ionised gas is deduced to be $\lesssim 10^6$ $M_\odot$. It was thought that most of the gas has been used up in very efficient star formation (Section III). This suggestion now seems inconsistent with colours of ellipticals (Faber and Gallagher 1976).

A number of surveys of early-type galaxies have been made for neutral hydrogen emission at 21 cm (Gallagher 1972, Balkowski *et al.* 1972, Bottinelli *et al.* 1973, Gallagher *et al.* 1975, Huchtmeier *et al.* 1975, 1977, Shostak *et al.* 1977, Balick *et al.* 1976, Knapp *et al.* 1977, Bieging and Biermann 1977, Knapp *et al.* 1978a, Krumm and Salpeter 1978a,b). There were two major motivations for these searches. Firstly, there is a well established correlation between neutral hydrogen content and morphological type for late-type galaxies (Roberts 1972); and it was hoped to extend the correlation to earlier types. Secondly, very conservative estimates of mass loss in ellipticals 0.1 - 1 $M_\odot\ yr^{-1}$ (Faber and Gallagher 1976) indicated that the HI should be readily detectable using available radio telescopes if the gas is retained in the galaxy in the form of atomic hydrogen.

The major results of these surveys are that ellipticals contain at most 0.1 percent of HI by mass with the average value of the total hydrogen mass to photographic luminosity $< M_{HI}/L_{pg} > < 0.009$. The SO's have $< M_{HI}/L_{pg} >$ varying between 0.001 - 0.05 with a mean of 0.015. A tentative correlation of gas content (M_{HI}/L_{pg}) with colour (B-V) was found by Balick *et al.* (1976) for SO's; but no such correlation was found for ellipticals (Knapp *et al.* 1978a).

In evaluating the conclusions of these surveys there are some important points to note. Firstly, baseline curvature and other instrumental effects present in the observed spectra indicate that, for massive galaxies with large velocity dispersion $\Delta V \propto L_{pg}^{1/4}$ (Faber and Jackson 1972), the line widths may be too large for a detection to be made, even if hydrogen masses $\gtrsim 10^9\ M_\odot$ were present. Secondly, great care must be taken to avoid mistaking hydrogen emission in neighbouring spirals for HI in an elliptical. Such caution was necessary in the three distinct cases of an early-type galaxy (2 ellipticals and an SO) paired with a spiral for which HI has been detected in the early-type galaxy of each interacting pair (NGC 3226/7, NGC 7332-9, NGC 1052/1042). Thirdly, in estimating upper limits to HI mass, a value of the velocity dispersion depending on the galaxy luminosity such as $\Delta V \propto (L_{pg}/L_\odot)^{1/4}$ (Knapp *et al.* 1978a) rather than a constant $\Delta V = 300\ km\ s^{-1}$, should be used.

B. Specific Objects

To date, after some extensive searches, there are 9 elliptical galaxies with HI detected. We shall consider these in some detail (Table 1), as well as 2 SO's of special interest.

Table 1. Properties of Ellipticals with detected HI

Galaxy	Type	$\Delta V (km\ s^{-1})$	$\frac{M_{HI}}{10^8 M_\odot}$	$\frac{M_{HI}}{L}$ $\frac{M_\odot}{L_\odot}$	[OII]	Radio (mJy)	Remarks
1052	E2	470	11	0.03	✓	460	
2974	E4	320	2.9	0.015	✓	< 25	
3226	E2p	510	> 2.9	0.02	✓	< 10	
3904	E2	480	7.5	0.06	x	< 10	Recent SN
3962	E2	460	8.4	0.05	✓	< 10	
4105	E3	380	5.2	0.05	✓	< 10	
4278	E1	470	2.5	0.05	✓	480	
4636	E0	380	8.2	0.04	✓	50	Recent SN
5846	E0-1	480	3.3	0.01	✓	17	

1. NGC 1052. - This E2 galaxy has strong optical emission and a powerful compact non-thermal radio source (Fosbury *et al.* 1978) with $M_{HI}/L = 0.03\ M_\odot/L_\odot$ and a mass in HI of $\sim 11 \times 10^8\ M_\odot$ (Knapp *et al.* 1978c, Reif *et al.* 1978). There seems to be a bridge connecting it with the spiral galaxy NGC 1042.

2. NGC 2974 . - This is an isolated E4 galaxy with $M_{HI}/L = 0.015$ Bottinelli and Gougenheim 1978b).

3. NGC 3226/3227.- This is a binary pair consisting of the strong emission line galaxy NGC 3226 type E2p and the broad emission line type 1 Seyfert Sa p (NGC 3227). The mass of neutral hydrogen is not centred on NGC 3227. The profiles have wide wings suggesting the presence of gas motion; and there is a feature at 1270 km s^{-1} that may be associated with both galaxies, possibly due to an arm connecting

them (Knapp *et al.* 1978).

4. NGC 3904. - This E2 galaxy has $M_{HI}/L = 0.06$, $\Delta V = 480 \pm 100$ km s^{-1}, and $M_{HI} = 7.5 \times 10^8 M_\odot$ (Bottinelli and Gougenheim 1977) but is the only one of the nine hydrogen-rich ellipticals not to show [OII].

5. NGC 3962. - This E2 galaxy also has a large scale distribution of neutral hydrogen similar to that found in NGC 4636 (Bottinelli and Gouguenheim 1978b) with a hydrogen extent 1.7 times larger than the optical galaxy.

6. NGC 4105. - Again, the detection of Huchtmeier *et al.* (1977) is confirmed by Bottinelli and Gouguenheim (1978b). This E3 galaxy also has a double horn-shaped profile with $M_{HI}/L \sim 0.05$. It is a member of a group, it could be an interacting pair with NGC 4104, and there is distinct activity in the nucleus.

7. NGC 4278. - A normal E1 galaxy with a strong central non-thermal radio continuum source and strong optical emission in [OII] λ 3727. The neutral hydrogen extends beyond the optical galaxy and is observed to have an elliptical distribution, with an angle of $\sim 50^\circ$ between the minor axes of the galaxy and the ellipse (Gallagher *et al.* 1977, Bottinelli and Gouguenheim 1977, Knapp *et al.* 1978). The rotation curve is flat at > 2 arcmin from the centre, giving evidence for a massive halo with $M/L \approx 26(M_\odot/L_\odot)$. The line profiles are gaussian, not the steep-sided profiles typical of disk galaxies.

8. NGC 4636. - This E0 galaxy has extended HI emission to approximately $\sim$ 1.3 times the optical size of the galaxy with a hydrogen mass of $\sim 8 \times 10^8 M_\odot$ and $M_{HI}/L = 0.04 M_\odot/L_\odot$ (Bottinelli and Gouguenheim 1977b, 1978a, Knapp *et al.* 1978c). It is not yet established if the HI is in the form of a disk or an ellipsoidal distribution as in NGC 4278.

9. NGC 5846. - The detection of Huchtmeier *et al.* (1977) for this E0 1 galaxy has been confirmed by Bottinclli and Gougucnheim (1978b). It has a double horn-shaped profile $M_{HI}/L \sim 0.047$.

10. NGC 7332. - This is a genuine edge-on S0 in a pair with the Sb spiral 7339 and a hydrogen mass of $\sim 1.5 \times 10^8 M_\odot$ (Knapp *et al.* 1978a).

11. NGC 2685. - This is an S0 p which appears to have recently accreted or captured some HI. A detailed discussion is given by Shane (1978).

III. GAS REMOVAL

The best conservative estimate for mass lost rate in ellipticals is $\sim$ 0.015 $(L/10^9 L_\odot)$ $M_\odot$ yr^{-1} (Faber and Gallagher 1974). For a constant mass loss rate over $\sim$ 4 x 10^9 yr (the timescale for which ellipticals may have undergone no significant change (Wilkinson and Oke 1978)) Knapp *et al.* (1978) find that the minimum estimate of $M_{HI} \gtrsim 4.4 \times 10^7$ $(L/10^9 L_\odot)$ $M_\odot$. Knapp *et al.* (1978a) demonstrate that in many cases this minimum lies well above the observed value. Note that in NGC 4278 it *agrees* with the correct value. The mass loss rate estimate is very much a minimum since, as discussed by Faber and Gallagher (1974), if a Salpeter luminosity function is used, a typical elliptical has a total mass loss in its lifetime of $\sim 6 \times 10^{11}$ $M_\odot$.

Let us summarize theories of gas removal: -

Star Formation

If star formation were to use up the gas, then, with a star formation rate proportional to a positive power of the density, elliptical galaxies should be bluer toward the cores (Larson 1974). There is now good observational evidence that ellipticals tend to be redder toward their central regions (Faber 1977). Bursts of star formation can be ruled out. In order to keep the gas level acceptably low, the timescale between bursts would have to be so short that bursts would not have time to decay and the effects of star formation in ellipticals would still be visible in integrated colours (Faber and Gallagher 1976).

Evaporation

As discussed by Cowie and Songaila (1977), Cowie and McKee (1977), and McKee and Cowie (1977), this process is relevant when the galaxy is embedded in a hot intergalactic medium such as that typical of an X-ray emitting cluster. The mechanism is essentially due to electron thermal conduction raising the temperature of the gas in the galaxy so that it is no longer gravitationally bound and then flows into the intergalactic medium.

The process only occurs for galaxies in X-ray emitting clusters. Moreover, it seems to be too efficient in its present form; for it is hard to see why any galaxies would contain gas in rich clusters. Obviously, an inhibiting mechanism is necessary and magnetic field effects are the obvious candidates since the ratio of the thermal conductivity across the field to that along the field is $\sim (r_e/l_f)^3$ where r_e is the gyroradius and l_f is the mean free path for collisions.

Winds

As analysed in detail by Matthews and Baker (1971) and Faber and Gallagher (1976), the mass lost from stars in elliptical galaxies could be forced out in a supernovae driven wind. It is interesting to note here that such a wind may also be present in our own galaxy, as can be derived from the 3-phase model of its interstellar medium (McKee and Ostriker 1978). Gisler (1976) has demonstrated for elliptical galaxies, however, that if the mixing of thermal energy between the hot supernova phase and the cooler phase of mass lost from the evolved stars has an efficiency of ε, then there is an upper limit to the mass of a galaxy that can drive a wind given by $M_{upper} \sim (\varepsilon)^3 \times 10^{12}\ M_\odot$. Thus, particularly if galaxies have massive halos, it may be difficult in many cases to drive a wind. Such is probably the case with M87 which recent X-ray observations indicate has a massive static halo of hot gas. The amount of mass required in M87's halo is too large to be explained by an outflowing wind (Matthews 1978a,b).

These winds can be thermally unstable if the outflow time of a fluid element in the wind is longer than its cooling time. For the lower mass galaxies whose effective stellar temperature $T_* = GM/RM_*$ is $\sim 10^5$ K (with a stellar velocity dispersion of 70 km s^{-1}) which is near the maximum of the cooling curve, then the wind may be inhibited by cooling. Two nearby galaxies in which this effect may be present are NGC 185 and NGC 205.

The hot low density gas of winds is not directly observable. Winds may be detected in the future from infrared and ultraviolet observations of their dust component. Quasar absorption lines could also supply information on winds, particularly if extended clumpy gaseous halos exist around the galaxies (Wolfe 1978, Boksenberg and Sargent 1978). Indirect evidence for winds is given by the 1 keV and 7 keV Fe lines with nearly solar abundance found in the intracluster medium (Serlemitsos *et al.* 1977, deYoung 1978).

Sweeping

The fourth mechanism proposed for gas removal from ellipticals is the sweeping action of an external medium with density ρ_e and a galaxy velocity v_g having a ram pressure $\rho_e v_g^2$ (Gisler 1976, 1978; Lea and deYoung 1976; Gunn and Gott 1972). Sweeping has a strict limitation: the mass injection rate into the interstellar medium of ellipticals is a fundamental parameter, so that, if this rate is high, stripping of the gas can never take place. But once the mass injection rate drops below a certain value, a galaxy then becomes vulnerable to sweeping. Gisler (1978) gives typical curves showing the time dependence of the mass injection rate. This inhibited stripping theory also has an important consequence for the evolution of radio sources (see Section V).

The theory offers a simple explanation of Butcher's and Oemler's recent observations (1978) that there is a rapid increase in the number of blue galaxies in clusters at redshifts of ~ 0.4. The increase is thought to be due to the number of spirals which, at a later time, will have become SO's due to gas removal. Melnick and Sargent (1977) show that SO evolution in clusters is consistent with either sweeping or evaporation, since there is a strong increase of the percentage of SO's toward the centre of X -ray clusters where both mechanisms would tend to be more efficient. Moreover, the colours of SO's are consistent with the hypothesis that they are normal spirals with their gas removed (Biermann and Tinsley 1975). Van den Bergh (1975) has argued that there is a class of spirals parallelling the nominal Hubble sequence termed "anaemic" which are gas deficient. Supporting evidence for this has been presented by Sullivan and Johnson (1978) who have found anomalously low values for M_{HI}/L_{pg} for spirals in clusters. Furthermore, Gisler (1978) has made an extensive survey of emission line galaxies. He has shown that for all types, not only E and SO galaxies, emission lines are more likely to be present in galaxies outside, rather than inside, rich clusters and concludes that this is evidence for sweeping or evaporation mechanisms.

If none of the foregoing mechanisms function to remove gas from an early-type galaxy, it may be that gas is stored in a different, unobservable form. Yet, none of the storage mechanisms examined by Faber and Gallagher (1976) seem viable, with one possible exception. Recent observations of the Sombrero galaxy 4594 (Bajaja 1978, Faber *et al.* 1977) provide good circumstantial evidence via the gas-dust correlation for believing that a significant amount of gas may be in molecular form. CO observations may prove useful in this case.

IV. ELLIPTICALS WITH H I: ACCRETION

It now remains to address the problem of the 9 ellipticals observed to contain HI. Why do *these* ellipticals have detectable gas and others not?

The most studied case is NGC 4278. The gas appears to be in a rotating disk or ellipsoid with its minor axis at an inclination of 50° from that of the optical galaxy. The timescale for alignment of the disk and galaxy axes appears to be very short, $\sim 10^9$ years. Two possible explanations are the outflow of gas from the nucleus, or tidal interaction with its nearest neighbours (NGC 4283 and NGC 4286). A third suggestion is that the cloud may have been recently *accreted* from the intergalactic medium.

In considering the argument for accretion as a more common phenomenon in ellipticals, let us first review the nature of galactic hydrogen clouds as potential sources of gas to feed the accretion process.

Intergalactic Hydrogen Clouds

Matthewson *et al.* (1975) have found evidence for HI clouds in the Sculptor Group with column densities of $\sim 10^{19}$ cm^{-2}. Another candidate for a cloud associated with this group has been found by

Cesarsky *et al.* (1977). Clouds, such as Wright's object near M33 (Wright 1974) or the Magellanic stream, have been found associated with galaxies. The interpretation of their associations is still somewhat uncertain. Haynes and Roberts (1978) have proposed that the clouds found by Matthewson *et al.* are not in the Sculptor group. Because of the large velocity difference between Wright's M33 cloud and M33, it could be a normal high-velocity cloud not associated with M33 at all (Oort 1978, private communication). If the extra-galactic interpretation is accepted, typical inferred cloud masses are $10^8 - 10^9$ $M_\odot$ bringing them interestingly close to observed values for the HI masses detected in ellipticals.

Limits on isolated HI clouds not in groups of galaxies are given by Roberts and Steigerwald (1977), who searched for clouds both in emission and absorption at 21 cm. The absorption measurements taken against radio galaxies gave a limit $n\,\sigma = 2.2 \times 10^{-3}$ Mpc^{-2} where σ is a cloud cross section and n is the number density of clouds. The emission data gave an upper limit of $n \sim 0.08$ Mpc^{-3} for clouds not in groups with masses $M \gtrsim 10^9$ $M_\odot$.

Extensive searches of groups of galaxies by Lo and Sargent (1978) have revealed a number of clouds of column density $\sim 10^{19}$ cm^{-2} with masses typically $\sim 10^8$ $M_\odot$. In general, these clouds appear to have an associated central blue stellar component with M_{HI}/L_{pg} the largest known value to date. These clouds could be primordial in origin with intermittent star formation occuring in their central regions, possibly induced by shocks developed in galaxy-cloud collisions. Alternatively, the clouds may be remnants of the tidal stripping of gas during galaxy collisions. Substantial mass of gas $\sim 10^8$ $M_\odot$ can be stripped from spirals in close low velocity encounters (Icke 1978, private communication), in which case the gas would be heavy element enriched.

Independent of their origin, individual HI clouds have an upper limit on their mass, since this must not exceed the Jeans' mass. A lower limit is given by the criterion that they must not evaporate in a Hubble time. Indeed, extensive observational searches have failed to detect any such clouds in rich X-ray emitting clusters because the high gas temperature in X-ray clusters makes the clouds evaporate on short timescales.

The limits on the mass of individual clouds can be written quantitatively as:

$$5 \times 10^7 T_6^{15/4} T_4^{1/2} p^{-1/2}\ M_\odot < M < M_J \equiv 10^9\ T_4^{\,2}\ p^{-1/2}\ M_\odot$$

where p is the pressure (cm^{-3} K) in both the clouds and intragroup medium. T_4 is the cloud temperature and T_6 the temperature of the intragroup medium.

An upper limit on the number density N of clouds of radius R, velocity v, in a typical group of galaxies can be obtained because cloud-cloud collisions are likely to be disruptive (Stone 1971). The condition that the cloud-cloud collision time exceeds a Hubble time means that

$$N \lesssim 10^4 \left(\frac{R}{10\ \mathrm{kpc}}\right)^{-2} \left(\frac{v}{100\ \mathrm{km\ s^{-1}}}\right) \mathrm{Mpc}^{-3}.$$

As discussed by Sargent (1977) and Wolfe (1978), QSO absorption lines may give us additional information on intergalactic HI clouds. The results of a search of the redshift systems of 19 QSO's are consistent with the hypothesis that QSO absorption lines are due to extended galactic halos with characteristic radius of $\sim$ 100 kpc (Sargent 1977). The line widths 150 km s^{-1} and splittings 100 km sec^{-1} of the lines support the idea that extended clumpy gaseous halos can produce the required absorption characteristics. Wolfe (1978) has discussed in detail the properties of the state of gas in gaseous halos that can be inferred from absorption lines in NGC 3067 and AO 0235 + 164.

Accretion

In summary, it seems that cloud masses in the necessary range could be available as sources for possible accretion processes, particularly in groups of galaxies. We can now consider the accretion of an isolated HI cloud by an elliptical or SO galaxy which has a wind or a tenuous atmosphere. (For further details see Silk and Norman 1978).

The condition for infall of a spherical cloud of density ρ, radius R captured by a spherical galaxy with a supersonic wind velocity v_w is that the cloud's column density must be greater than a certain critical value given by

$$(\rho R)_{crit} = \frac{3}{16\ \pi} \left(\frac{\dot{M}}{M} \frac{v_w}{G}\right) = 10^{-6} \left(\frac{v_w}{100\ \mathrm{km\ s^{-1}}}\right)\left(\frac{M/\dot{M}}{10^{12}\mathrm{yr}}\right)^{-1} \mathrm{g\ cm}^{-2},$$

where $\dot{M}$ is the mass loss rate and M is the total mass of the galaxy. A similar condition holds for the column density of a shell of gas surrounding the galaxy when the gas is prevented from falling in by the action of an outflowing wind.

The calculations of cloud infall assumes an approximate pressure balance with the ambient medium or wind. Clouds which cannot cool will be compressed adiabatically with $T \propto \rho^{2/3}$ and will fall until they reach the radius at which buoyancy forces become significant. At this point it is expected that metal enrichment can occur, the temperature will drop to 10^4 K, and star formation will result. Clouds that can cool will remain isothermal at $\sim 10^4$ K. As they fall in and the pressure rises, star formation will also be initiated. Thus, dependent on their cooling capabilities, clouds will form stars in the outer or inner parts of a galaxy.

Enhancement of heavy element abundance gradients can be qualitatively understood in this picture. As metal enrichment of a cloud initiates star formation, the greater the abundance gradient, the greater the star formation gradient.

A consequence of the interaction between outflowing winds and infalling clouds is a feedback process generating bursts of star formation (Silk and Norman 1978). Accretion of clouds will lead to star formation and possibly to even more violent activity in the nucleus (see Section V).

Assuming a reasonable mass function for this star formation, the supernova rate will increase and amplify the wind strength, which in turn will inhibit the accretion. An observable result of the burst would be to drive out a shell or bubble of gas in a shock wave.

The timescale for a typical burst is given by the ratio of the total mass in stars to the initial mass loss rate in the burst which depends on the assumed mass function and supernova rate and is typically $\sim 10^{7-8}$ yr.

The period for the cloud or shell to reaccrete, or for some other HI refuelling process to occur, is on the order of $\sim 10^{9-10}$ yr.

It seems that a wind accretion interaction-model would also be applicable to Seyfert galaxies. They appear to belong to the class of early-type SO or Sa galaxies with a dominant spheroidal component (de Vaucouleurs 1975; see however Adams 1977). The accretion phenomenon in the Seyfert case would not be due to intergalactic clouds, but to the HI supply fuelled in some way from the gas-rich disk. The characteristic timescale for bursts of star formation in the previous model corresponds with the active phase of Seyfert nuclei. The quiescent phase of Seyferts, corresponding to the time between bursts, is the refuelling time. These two phases should have timescales with a ratio of ~ 0.01 to account for the 1 percent of spirals in the Seyfert class. The evidence given by Bieging and Biermann (1977) and by Heckman *et al.* (1978) for ring and shell structures in SO's and Seyferts is consistent with the bubble model presented here as the result of a burst driving out the accreting matter.

We may be seeing the very same process of explosion/accretion interaction in the "active" galaxies, M82 and Cen A, as well.

The formation of flattened HI distributions can also be taken into account by the accretion process if obviously discrete clouds have angular momentum. In that event, they will spiral toward the nucleus in nearly Keplerian orbits of radius r, since the ratio of the spiralling-in time to the rotation period is large. If friction with the wind is the dominant mechanism for a cloud to lose its angular momentum, then the radial infall velocity v_r is given by:

$$v_r = 2\pi \left(\frac{R}{r}\right)^2 \left(\frac{\dot{M}}{M}\right) \left(\frac{GM}{v_w^2 r}\right)^{1/2} r.$$

As the wind strength increases, the radial infall velocity increases. This relation demonstrates a positive feedback on fuelling activity in the central core until the wind strength rises above the critical value for a given column density, the burst shuts off, and the infalling clouds are driven outwards. The problem of how clouds lose their angular momentum and spiral into the nucleus in a sufficiently short time to fuel activity in these cores, seems one of the central theoretical problems faced by models which invoke accretion as the mechanism for powering active galactic nuclei (Gunn 1978).

V. EVOLUTION OF ELLIPTICALS AND SO'S: RADIO SOURCES

Although early-type galaxies appear to be gas-deficient in the present epoch and although there is little evidence of star formation in ellipticals, the picture may have been quite different at a look-back time of 1/3 of a Hubble time ($z = 0.5$). In a recent survey of galaxies in two rich clusters with mean cluster redshifts of $z = 0.46$ and $z = 0.34$, Butcher and Oemler (1978) found that 30 to 50 percent of these galaxies had colours consistent with those of spiral galaxies, contrasting with the old population ellipticals and SO's that constitute the Coma cluster. The results of Kron *et al.* (1978), Spinrad (1977), Kristian *et al.* (1978), and of Hawkins and Reddish (1975) confirm this finding. One possible interpretation is that SO's had not yet formed by $z \sim 0.5$ and were still in their normal spiral phase; so that it is only in the last 1/3 of a Hubble time that gas has been removed from them and the gas-free SO's formed.

Additional information on this problem comes from recent work by de Ruiter *et al.* (1978). They find indications that the underlying galaxies associated with non-thermal radio sources may be bluer by about a magnitude in B-R at redshifts of about $z = 0.6$. The blue excess is with respect to zero evolution, even if an M87 type k-correction is used. It remains to be seen whether the blueness is due to a central continuum source, or if it is evidence for star formation.

Much of the above evidence then points to the relatively recent operation ($z < 0.5$) of gas removal mechanisms. This may have important consequences for our understanding of non-thermal radio sources which show a steep increase in comoving space density with increasing redshift.

Gisler (1976) has been able to model this relationship between gas content and activity in the nucleus using the straightforward assumption that the injection rate of mass lost from the stars and retained by (i.e. not stripped from) the galaxy must exceed a certain critical value $\dot{M}_{crit}$ in order to force a galactic nucleus into activity. More specifically, if galaxies have a Maxwellian velocity distribution and a given time-dependent mass injection rate, then it is possible to formulate a critical velocity for the galaxy as a function of the mass injection rate. Galaxies moving above this critical velocity will be unable to maintain a sufficiently high density atmosphere and correspondingly short cooling time for the nucleus to be fuelled by accretion.

If one adjusts the mass injection rate used to fuel the nucleus to be very roughly 1 percent of the mass lost from the stars then the characteristic timescale can be matched to model both the evolution of the space density of radio sources and the evolution of galaxies due to gas removal.

In a more general form, the theory should account for the presence of powerful radio sources in nearby rich clusters. Such sources often appear to be the dominant giant elliptical galaxies at the centre of rich clusters. These giant ellipticals are probably formed from multiple mergers (Ostriker 1977; Roos and Norman 1978). They have low velocities through the intra-cluster medium; and recent X-ray observations (Gorenstein *et al.* 1977) indicate that they may have extended static atmospheres, as discussed by Matthews (1978a,b). The gravitational confinement of such extensive hot atmospheres implies the existence of a super-massive halo in the particular case of M87, which has been estimated at 10^{14} $M_{\odot}$. It seems that, in such important cases as M87 and NGC 1275, gas removal mechanisms must be inefficient. In fact, the power in these sources is most probably derived from the infall of cool gas from the extended atmospheres (Rubin *et al.* 1977).

VI. CONCLUSIONS

To date, there are 9 exceptions to what seems to be a general rule for the absence of gas from elliptical galaxies. Evidently, some mechanism is at work to remove gas from the majority of ellipticals: either by winds, or sweeping, or evaporation, or by star formation. These mechanisms may have been efficient only for a lookback time of about 1/3 of a Hubble time. The S0's give the appearence of a true transition class between ellipticals and normal spirals with respect to the evolution of their gas content. Furthermore, this steep evolution in the gas content of early-type galaxies can be plausibly related to the comoving space density evolution of radio sources. The assumption is that an interstellar medium with a density above a critical value, and thus a sufficiently short cooling time, is a necessary condition for the fuelling of active galactic nuclei.

Future high resolution observations of HI in early-type galaxies will map the flow and distribution of either infalling clouds or gas transfer between galaxies. When such observations are coupled with both radio continuum and optical studies of the nuclei of such galaxies as NGC 1052 (Fosbury *et al.* 1978), we can expect a considerable increase in our understanding of the fuelling process of active nuclei. An additional and, for many, extremely significant measurement will be of the rotation curves for gas far outside the optical galaxy, as has been done for NGC 4278, gathering evidence for or against the existence of massive dark halos in ellipticals.

From all the recent studies then, it seems beyond question that gas in early-type galaxies, both by its presence even in small quantities and by its apparent absence from most ellipticals, is of special importance.

ACKNOWLEDGEMENTS

It is a pleasure to acknowledge helpful discussions and/or correspondence with many colleagues in particular Drs. Bottinelli, Eastman, Gouguenheim, Hummel and Shane. I have benefited greatly from collaboration with Dr. J. Silk on this subject by way of a travel grant from the NATO Scientific Affairs Division.

REFERENCES

Adams, T.F.: 1977, Astrophys. J. Suppl. Ser. 33, 19.
Bajaja, S.: 1978, preprint.
Balick, B., Faber, S.M., and Gallagher, J.S.: 1976, Astrophys. J. 209, 710.
Balkowski, C., Bottinelli, L., Gouguenheim, L., and Heidmann, J.: 1972, Astron. Astrophys. 21, 303.
Bieging, J.H. and Biermann, P.: 1977, Astron. Astrophys. 60, 361.
Biermann, P. and Tinsley, B.M.: 1975, Astron. Astrophys. 41, 441.
Boksenberg, A. and Sargent, W.L.W.: 1978, Astrophys. J. 220, 42.
Bottinelli, L. and Gouguenheim, L.: 1977a, Astron. Astrophys. 54, 641.
Bottinelli, L. and Gouguenheim, L.: 1977b, Astron. Astrophys. 60, L23.
Bottinelli, L. and Gouguenheim, L.: 1978a, Astron. Astrophys. 64, L3.
Bottinelli, L. and Gouguenheim, L.: 1978b, Astron. Astrophys. (in press).
Bottinelli, L., Gouguenheim, L., and Heidmann, J.: 1973, Astron. Astrophys. 25, 451.
Butcher, H. and Oemler, A.: 1978, Astrophys. J. 219, 18.
Cesarsky, D.A., Falgarone, E.G., and Lequeux, J.: 1977, Astron. Astrophys. 59, L5.
Cowie, L.L. and McKee, C.F.: 1977, Astrophys. J. 211, 135.
Cowie, L.L. and Songaila, A.: 1977, Nature 266, 501.
de Vaucouleurs, G.: 1975, Astrophys. J. Suppl. Ser. 29, 193.
deYoung, D.S.: 1978, Astrophys. J. 223, 47.
Ekers, R.D.: 1978, in IAU Symposium No. 77, *Structure and Properties of Nearby Galaxies*, p. 49, Eds. E.M. Berkhuisen and R. Wielebinski, D. Reidel Publ. Co., Dordrecht, Holland.
Faber, S.M., Balick, B., Gallagher, J.S. and Knapp, G.R.: 1977, Astrophys. J. 214, 383.
Faber, S.M. and Gallagher, J.S.: 1976, Astrophys. J. 204, 365.
Faber, S. and Jackson, R.: 1976, Astrophys. J. 204, 668.
Fosbury, R.A.E., Mebold, U., Goss, W.M., and Dopita, M.A.: 1978, Mon. Not. R. Astron. Soc. 183, 549.
Gallagher, J.S.: 1972, Astron. J. 77, 568.
Gallagher, J.S., Faber, S.M., and Balick, B.: 1975, Astrophys. J. 202, 7.
Gallagher, J.S., Knapp, G.R., Faber, S.M., and Balick, B.: 1977, Astrophys. J. 215, 463.
Gisler, G.R.: 1976, Astron. Astrophys. 51, 137.
Gisler, G.R.: 1978, Mon. Not. R. Astron. Soc. 183, 633.
Gisler, G.R.: 1978, preprint.
Gorenstein, P., Topka, K., Tucker, W., and Harnden, F.R.: 1977, Astrophys. J. 216, L95.
Gunn, J.E.: 1978, preprint.
Gunn, J.E. and Gott III, J.R.: 1972, Astrophys. J. 176, 1.
Gunn, J.E. and Oke, J.B.: 1975, Astrophys. J. 195, 255.
Hawkins, M.R.S. and Reddish, V.C.: 1975, Nature 257, 772.
Haynes, M., and Roberts, M.S.: 1978, Bull. American Astron. Soc. 9, 584.
Heckmann, T.M., Balick, B., and Sullivan, W.T.: 1978, Astrophys. J. 224.
Huchtmeier, W.K., Tamman, G.A., and Wendker, H.J.: 1975, Astron. Astrophys. 42, 205.

Huchtmeier, W.K., Tamman, G.A., and Wendker, H.J.: 1977, Astron. Astrophys. 57, 313.
Knapp, G.R., Faber, S.M., and Gallagher, J.S.: 1978c, Astron. J. 83, 11.
Knapp, G.R., Gallagher, J.S., and Faber, S.M.: 1978b, Astron. J. 83, 139.
Knapp, G.R., Gallagher, J.S., Faber, S.M., and Balick, B.: 1977, Astron. J. 82, 106.
Knapp, G.R., Kerr, F.J., and Williams, B.A.: 1978a, Astrophys. J. 222, 800.
Kristian, J., Sandage, A., and Westphal, J.A.: 1978, Astrophys. J. 221, 383.
Kron, R.G., Spinrad, H., and King, I.R.: 1977, Astrophys. J. 217, 951.
Krumm, N., and Salpeter, E.E.: 1978a,b, preprint.
Larson, R.B.: 1974, Mon. Not. R. astron. Soc. 166, 585.
Lea, S.M. and deYoung, D.S.: 1976, Astrophys. J. 210, 647.
Lo, K.Y. and Sargent, W.L.W.: 1978, preprint.
Mathews, W.G.: 1978a, Astrophys. J. 219, 408.
Mathews, W.G.: 1978b, Astrophys. J. 219, 413.
Mathews, W.G. and Baker, J.C.: 1971, Astrophys. J. 170, 241.
Mathewson, D.S., Cleary, M.N., and Murray, J.D.: 1975, Astrophys. J. 195, L97.
McKee, C.F. and Cowie, L.L.: 1977, Astrophys. J. 215, 213.
McKee, C.F. and Ostriker, J.P.: 1978, Astrophys. J. 219, 292.
Melnick, J. and Sargent, W.L.W.: 1977, Astrophys. J. 215, 401.
Minkowski, R. and Osterbrock, D.E.: 1959, Astrophys. J. 129, 583.
Morgan, W.W. and Osterbrock, D.E.: 1969, Astron. J. 74, 515.
Osterbrock, D.E.: 1960, Astrophys. J. 132, 325.
Ostriker, J.P.: 1977, *Evolution of Galaxies and Stellar Populations*, p. 369, Eds. Tinsley, B.M. and Larson. R.B., Yale Univ. Obs.
Reif, K., Mebold, U., and Goss, W.M.: 1978, Astron. Astrophys. 67, L1.
Roberts, M.S. and Steigerwald, D.G.: 1977, Astrophys. J. 217, 883.
Roos, N. and Norman, C.A.: 1978, preprint.
Rubin, V.C., Ford, W.K., Peterson, C.J., and Oort, J.H.: 1977, Astrophys. J. 211, 693.
de Ruiter, H., Katgert, P., and van der Laan, H.: 1978, preprint.
Sargent, W.L.W.: 1977, *Evolution of Galaxies and Stellar Populations*, p. 427, Eds. Tinsley, B.M. and Larson, R.B., Yale Univ. Obs.
Serlemitsos, P.J., Smith, B.W., Boldt, E.A., Holt, S.S., and Swank,J.H.: 1977, Astrophys. J. 211, L63.
Shane, W.: 1978, preprint.
Shostak, G.S., Roberts, M.S., and Peterson, S.D.: 1975, Astron. J. 80, 581.
Silk, J. and Norman, C.A.: 1978, in preparation.
Spinrad, H.: 1977, *Evolution of Galaxies and Stellar Populations*, 301, Eds. Tinsley, B.M. and Larson, R.B., Yale Univ. Obs.
Stone, M.E.: 1970, Astrophys. J. 159, 277.
Sullivan, W.T. and Johnson, P.E.: 1978, preprint.
van den Bergh, S.: 1975, Ann. Rev. Astron. Astrophys. 13, 217.
Wilkinson, A. and Oke, J.B.: 1978, Astrophys. J. 220, 376.
Wright, M.C.H.: 1974, Astron. Astrophys. 31, 317.
Wolfe, A.M.: 1978, NATO Advanced Study Institute on Energy Sources and Emission Mechanisms in Quasars (Cambridge).

RADIATIVE CONTINUUM SOURCES AND DENSE LINE REGION IN SEYFERT TYPE I GALAXIES AND BROAD LINE RADIO GALAXIES

J. Bergeron
European Southern Observatory

ABSTRACT

The radiative energy content of Seyfert I galaxies and broad line radio galaxies (BLRG) is reviewed. Most of the power is emitted at far infrared and hard X-ray energies. The variability of the continuum radiation and the life time of the active phase give some information on the source sizes. A lower limit to the size of the X-ray and soft γ-ray sources is given by the opacity to electron scattering and by the efficiency of the electron pair production. The broad line region (BLR) can be responsible for the observed keV absorption. It is also optically thick in the Lyman continuum. Its geometrical configuration is then derived. Free-free absorption in the BLR is very efficient at centimeter wavelengths and the BLR must not surround the compact radio source of the BLRG. The optical and UV absorption lines must arise from regions outside the BLR. The spatial coverage of the Hβ emitting region is large but that of the FeII region may be smaller and may greatly vary between galaxies.

I RADIATIVE ENERGY CONTENT

The infrared and X-ray observations of Seyfert I galaxies and BLRG are reviewed. Most of the radiative power is emitted in these energy ranges even for radio sources.

a. IR Emission

IR observations of Seyfert galaxies have been recently reviewed by Neugebauer (1978). A composite origin of the emission has been strongly favoured (see e.g. Khachikian and Weedman 1974): a galactic stellar component, a thermal component from heated dust and a non-thermal component. The assumption of a much weaker thermal component in Seyfert I galaxies than in Seyfert II galaxies, thus a lower ratio of 10 μ to UV fluxes (Neugebauer *et al.* 1976, Stein and Weedman

Bengt E. Westerlund (ed.), Stars and Star Systems, 67–80.

1976) is not supported by the recent more extensive data of Rieke (1978).

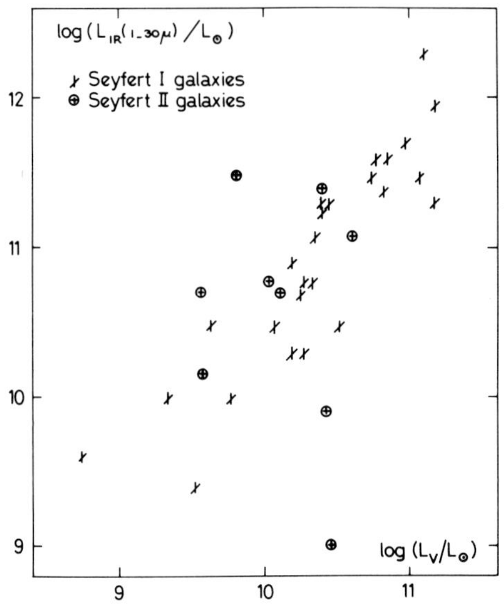

Fig. 1. IR luminosity versus optical luminosity for Seyfert I and Seyfert II galaxies.

The IR luminosity (using the extrapolation between 10 and 30 μ as given by Rieke 1978) is shown in Figure 1 as a function of the optical luminosity for both types of Seyfert galaxies. The Hubble constant is taken equal to 50 km sec^{-1} Mpc^{-1}. The scatter is very large for the small sample of Seyfert II galaxies (the low IR luminosity of 10^9 $L_\odot$ obtained for Mk 268 is a consequence of the spectral turn over at $3\ \mu < \lambda < 10\ \mu$ and the lack of observations at $\lambda > 10\ \mu$). Yet this class does not markedly differ from the Seyfert I class in the ratio of 10 μ to optical flux. The IR luminosity is roughly proportional to the optical luminosity, with $< L_{IR}\ (1 - 30\ \mu)/L_{opt} > \ \sim\ 4$ to 5 for Seyfert I galaxies. The power emitted by energy interval is on the average not much higher at 10 μ than in the optical.

The spectral distribution from 10 μ to 1.25 μ or to 3500 Å as given by Neugebauer *et al.* (1976) and Rieke (1978) greatly differ within objects of the same class. Spectra rising steeply at long wavelengths are found in both classes (NGC 1068 and Mk 231) though more frequently among Seyfert II galaxies. Variability of the IR emission of Seyfert I galaxies would be a crucial test for the assumption of non-thermal origin. The observations of Rieke (1978) do not confirm any variability at 10 μ with the exception of 3C 120 in which the observed slow change (a factor of 2 in 10 years) is consistent with variation time-scales in thermal sources. Small variations at 3.5 μ have been reported for NGC 4151

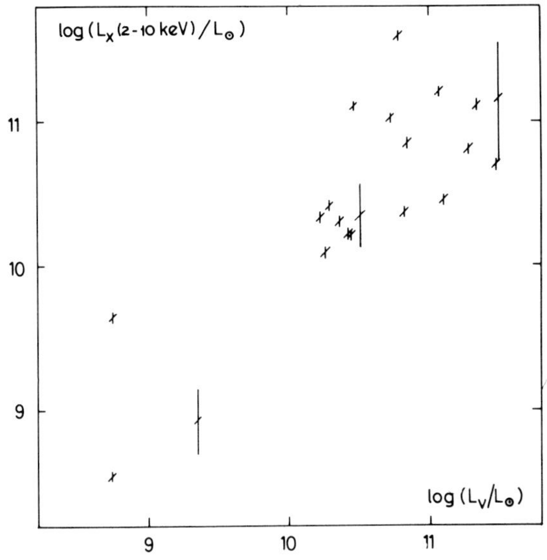

Fig. 2. X-ray luminosity versus optical luminosity for Seyfert I galaxies. The vertical bars account for the observed X-ray variation.

(Penston *et al.* 1974). This implies a combination of both thermal and non-thermal emission for Seyfert I galaxies, with a predominance of the thermal component at $\lambda \lesssim 10\ \mu$ and for some galaxies, but not all (e.g. Mk 231), a large thermal contribution at shorter wavelengths $\lambda \lesssim 3\ \mu$.

The IR luminosities, in the range 1 - 30 μ, given by Rieke (1978) are smaller than the total IR power for the stronger 10 μ sources. The turn over of the IR intensity occurs around 100 μ for NGC 1068 (Telesco *et al.* 1976) and 30 μ for NGC 4151 (Rieke and Low 1975). Yet even for NGC 1068 only half of the IR flux is emitted beyond 30 μ, thus L(1 - 30 μ) is not a too bad measure of the IR power.

b. X-ray Emission

All Seyfert I galaxies with visual magnitude V < 14.5 have been detected in the 2 - 10 keV energy range (Elvis *et al.* 1978, Tananbaum *et al.* 1978, Pounds 1978). The X-ray luminosity L(2 - 10 keV) to optical luminosity ratio is on average equal to unity as shown in Figure 2. The scatter is very large for both low and high luminosity galaxies.

Variability by factors of at least 2 has been detected for several Seyfert galaxies and BLRG (NGC 4151, MCG 8-11-11, 3C 120). The shorter time scales are one month for MCG 8-11-11 (Ward *et al.* 1977) and for

NGC 4151 3 days (Elvis 1976) and may be 10^3 sec (Tananbaum *et al.* 1978). These variation time scales are shorter than those observed in optical and near IR (Penston *et al.* 1974). The fastest optical continuum variations yet detected for NGC 4151 are of 15 percent over 15 days and of a factor of 2 over 4 months (Lyutyi 1973).

For a few galaxies spectral information is available (Stark *et al.* 1978 and preliminary results of HEAO-A2). The spectrum often shows considerable low energy absorption, revealing column densities N_H of the order of 10^{23} cm^{-2} (assuming solar abundances). More extensive data are available for NGC 4151, and N_H varies from 3.5×10^{22} to 1.8×10^{23} cm^{-2} (Barr *et al.* 1977). The shorter variability time scale for N_H seen by Ariel 5 is of the order of a few months (10^7 sec).

The 3 day variation of the X-ray flux observed for NGC 4151 could be explained either in terms of intensity variations or in terms of variation in the amount of absorbing material. The first assumption seems more likely. Indeed the elliptical galaxy Cen A exhibits intensity variations on a time scale of days without any variation in N_H (Mushotzky *et al.* 1978) and has roughly the same X-ray power and absorbing column density than NGC 4151.

All the sources observed with good spectral resolution show a hard spectrum $I_\nu \propto \nu^{-\alpha}$ with $\alpha < 1$. Observations by HEAO-A2 (Boldt 1978)confirm the hardness of the X-ray spectra until 80 keV. The values for α

Table 1

IR and X-ray spectral indices

Object	α_X	α_{IR}	Comments
Mk 279	-.3,.2	1.4	$N_{H,22} = 2$
Mk 506	.3		
Mk 509	.7	0.6	
NGC 3783	.75	1.3	
NGC 4151	.4,.55	1.45	var, $N_{H,22}$ = 3-20
NGC 5506	.6		$N_{H,22} = 3$
NGC 5548	.5	1.0	excess at low energy
NGC 6814	.5	2.2	
3C 120	.5	1.0	var, recent softening
3C 390.3	.7		$N_{H,22} = 5$

are given in table 1 for 10 Seyfert galaxies. The column densities known for even fewer cases are also presented. The low column density of NGC 5548 and maybe 3C 120 could most probably be explained by a non complete spatial coverage of the absorbing region. This will be further discussed in section III.

c. Hard X-ray and γ-ray Emission

NGC 4151 shows a flattening of the spectrum around 100 keV ($\alpha(E > 100$ keV$) \simeq 0$) and is a strong MeV source (Auriemma *et al.* 1978, Schönfelder 1978) with a break around 3 MeV. The X-ray luminosity in the range 2 - 100 keV is equal to the total IR luminosity. (This is also true on the average for all Seyfert I galaxies). The luminosity between 100 keV and 3 MeV derived from the observations by Schönfelder (1978) equals 8×10^{44} erg sec $^{-1}$ or 10^2 L(.33 - 1 μ) or 20 L(1 - 30μ). Only 13 percent of all Seyferts (Schönfelder 1978) can have similar MeV power and similar spectra as NGC 4151 to avoid a conflict with the observed diffuse cosmic γ-ray flux.

Observations in the 50 - 150 MeV range by Cos B (Bignami, private communication) yield a flux upper limit consistent with the slope $\alpha \sim 3$ measured between 3 and 10 MeV.

II. CONSTRAINTS ON THE SIZE OF THE CONTINUUM RADIATIVE SOURCE

The variations of the continuum emission provides one constraints on the maximum dimension of the radiative source. The shortest of them are observed at X-ray frequencies and are of the order of day(s) (NGC 4151, Mk 421). On the other hand lower limits on the source dimension could be obtained from the observations with some model assumptions.

a. Variability Versus Life-Time of the Active Phase

The minimum size of the source can be inferred from the amount of energy emitted during the active phase. The life time τ of the Seyfert phase should be at least 10^8 years if all spiral galaxies go through such an active phase (Weedmann 1977). The energy available is proportional to the rest mass of the source Mc^2 (if the mass of the source is steadily decreasing with time the argument remains valid assuming that in the coming 10^{10} years the Seyfert phenomena will be of similar amplitude than in the past). For a total radiative luminosity one gets

$$L \tau = \varepsilon Mc^2 , \qquad (1)$$

where ε is the efficiency. The Schwarzschild radius r_s associated to this mass is a lower limit of the radiative source size. If Δt is the shortest variation time scale observed one obtains

$$r_s/c \,\Delta t = (2G/c^4)\, \varepsilon^{-1}\, \tau (L/c\, \Delta t) < 1, \qquad (2)$$

for a spherically symmetric source and isotropic emission. The smallest variation time-scale compatible with a life-time of 10^8 years is equal to

$$\Delta t > \Delta t_{crit} = 20\ (\tau/10^8\ \text{yrs})\ (\varepsilon/0.1)^{-1}\ (L/10^{44}\text{erg sec}^{-1})\ \text{sec.} \quad (3)$$

The total luminosity is taken equal to the X-ray luminosity. The observed and computed variation time-scales are given in table 2 for a few galaxies, using the X-ray variability. For NGC 4151 Δt_{crit} is close to the shortest X-ray variations observed. If these variations were confirmed, the X-ray emission would barely be compatible with an isotropic emission around a black hole in the Schwarzschild metric case for an active phase length of 10^8 years. If not confirmed, the shortest variations observed are a few hundred times longer than Δ_{crit}. For more luminous objects with $L \simeq 10^{47}$ erg sec^{-1} (typically the quasar 3C 273) critical variations would have a time-scale of days.

Table 2

Variability of Active Nuclei

Galaxy	Δt_{obs} sec	$L(\Delta E_x)$ erg s^{-1}	ΔE_x keV	Δt_{crit} sec
NGC 4151	3×10^5	5×10^{42}	2 - 10	1
	"	1×10^{45}	2 - 10^4	200
	700	1×10^{45}	2 - 10^4	200
MGC 8-11-11	3×10^6	9×10^{43}	2 - 10	20
Mk 421	1×10^5	1×10^{45}	2 - 80	200

b. Electron Scattering

A large electron column density is required for thermal X-ray models. The thermal photons will be scattered by the electrons and the shortest variations observable are those taking place close to the outer layers within a length of optical depth unity.

The scattering optical depth is $\tau_T = \sigma_T\, n\, l$ where σ_T is the Thompson cross section. The density is related to the mass within the thermal source and this mass is defined by the amount of energy emitted during an observed flare of rise time Δt_{obs}. For a spherical symmetry this intensity variation will be observed if

$$\Delta t_{obs} \gtrsim \Delta t\ (\tau_T = 1) = (\sigma_T / 4 m_p\ c^4)\ \varepsilon^{-1}\ L = \tag{4}$$

$$= 120\ (\varepsilon/0.1)^{-1}\ (L/10^{44}\ \text{erg sec}^{-1})\ \text{sec.}$$

The luminosity which should be considered is that during the flare. For NGC 4151, no temporal variations are yet observed at hard X-ray energies. The 2 - 10 keV flare with a rise time of 700 sec and a peak luminosity of 4×10^{43} erg sec^{-1} (Tananbaum *et al.* 1978) is consistent with the electron scattering constraints, and the size of the emitting region must exceed $1.5 \times 10^{12}\ (\varepsilon/0.1)^{-1}$ cm.

c. Electron Pair Production

The photon density is large for the sizes inferred from the observed variations. If the γ-ray and X-ray photons are emitted in the same volume, the overall dimension of this high energy photon source must be large enough to avoid a large rate of electron pair production and the destruction of the γ-ray photons. The optical depth to electron pair production is

$$\tau_{e\pm} = \sigma n_x\, l\ , \tag{5}$$

where the cross section is energy dependent with a maximum $\sigma_{max} = \sigma_o/2 \simeq (3/16)\sigma_T$. The threshold of the reaction is reached for $s = E_\gamma\, E_x\ (1 - \cos\theta)\ /\ (2(mc^2)^2) = 1$, where $\theta = \pi/2$ for an isotropic emission.

For a spherically symmetric source of radius r and isotropic emission, the optical depth for γ-ray photons of energy E_γ may be written

$$\tau_{e\pm}\ (E_\gamma) = (\sigma_o / 8\ \pi m^3)\ r^{-1}\ L_{XO}\ (E_\gamma/mc^2)\ f(\alpha), \tag{6}$$

where $f(\alpha)$ is a slowly varying function of the intensity spectral index α(with $f(\alpha) = 0.4$ for $\alpha = 0.5$), and L_{XO} is the luminosity per energy interval defined at the reaction threshold energy

$$L_{XO} = n_{XO}\ E_{XO}^{\ 2}\ \ 4\pi r^2\ c. \tag{7}$$

The constraint $\tau_{e\pm}$ yields a lower limit on r or on Δt. When taking into account the energy dependence of the electron pair production cross section one obtains

$$\Delta t > 700 (L_{XO}/10^{44}\ \text{erg sec}^{-1})\ (E_\gamma/mc^2)\ \text{sec.} \tag{8}$$

For NGC 4151 a break in the γ-ray spectrum is observed at E_γ = 3MeV (Schönfelder 1978). The X-ray energy threshold is at 170 keV and $L_{XO} = 7 \times 10^{43}$ erg sec^{-1}, which gives $\Delta t > 3000$ sec, larger than the observed X-ray variation of 700 sec. Either these variations are not present or the X-ray and γ-ray sources do not coincide in space. The second alternative is not suggested by the smooth spectral distribution from the low X-ray up to the MeV energy range. Another alternative would be an anisotropic emission, as this increases the threshold energy, $E_\gamma = \sqrt{2}\, mc^2\, (1-\cos\theta)^{-1/2}$. For NGC 4151 the beaming angle required is $\theta < 20^\circ$.

The nearby quasar 3C 273 may have been detected by Cos B at energies of 100 MeV (Swanenburg *et al.* 1978). The 100 MeV emission is consistent with a power law extrapolation of the observed X-ray. The power law index is slightly smaller than unity and the luminosity per energy interval at 100 MeV is 2×10^{46} erg sec^{-1}. The inferred lower limit on Δt equals 9×10^6 sec or 3 months. Up to now variations on time-scale of one year have been observed either at radio, optical or X-ray energies (Neugebauer 1978 and Giacconi 1978). Forthcoming γ-ray observations will thus provide a very important clue to both the nature and the dimension of the active source.

III BROAD LINE REGION

The ionization structure and physical state of the broad line region, BLR, has been recently reviewed (Collin-Souffrin 1978 and Osterbrock 1978). We will present here some constraints on the size of the BLR and some line correlations not mentioned by the above authors.

a. Absorption of the Low Energy X-rays

Large column densities of the order of 10^{23} cm^{-2} (assuming normal abundances) have been observed for a few Seyfert I galaxies, the radio galaxies Cen A and 3C 390.3, and the narrow line galaxy NGC 5506.

Such large column densities are characteristic of the BLR, even with small filling factors, f. For the narrow line region, NLR, a large column density requires a large filling factor, thus a small overall size, r. The Balmer line intensities fix the value of $n^2 r^3 f$, the density n is given by the [OIII] lines (ratio of auroral to nebular lines) and the column density nrf, if known, determines f.

For NGC 4151 the NLR is optically resolved and of too low column density to account for the X-ray absorption. For NGC 5506, using the line intensities given by Wilson *et al.* (1976), we find $r(\mathrm{NLR}) \leq 20$ pc for $nrf = 3 \times 10^{22}$ cm^{-2}.

The column density for NGC 4151 varies on a time scale $\geq 10^7$ sec (section I). The observed values for N_H differ by a factor of 5. A small number of clouds or filaments are seen in the line of sight. Assuming a cloud column density of 5×10^{22} cm^{-2}, one obtains for the transverse velocity v_t a lower limit of $500\ (\Delta t/10^7\ \mathrm{sec})^{-1}$ $(n/10^8\ \mathrm{cm}^{-3})^{-1}$ km sec^{-1}. The density of the BLR cannot be much larger than 10^{11} cm^{-3} if the UV [CIII] lines are emitted in most of the BLR. Further, the transverse velocity cannot exceed the maximum velocity deduced from the emission line width, if the absorbing material and the BLR are a unique region. This yields a lower limit on Δt of 10^3 sec for $v_t = 5000$ km sec^{-1} (FWZI/2 $\simeq$ 6000 km sec^{-1} is given by Osterbrock 1971). If one considers the observed variation time scale for N_H and v_r given by FWZI/2, one gets $n = 10^9$ cm^{-3} and a filament size of 5×10^{13} cm.

The overall size and filling factor of the BLR could be determined if the amount of emitting material is known (this latter is given by the Balmer line intensities, assumed of radiative recombination origin. Although line transfer is dominant this should give values for n^2r^3f accurate within a factor of 3). Applied for NGC 4151 one derives for $nrf = 1 \times 10^{23}$ cm^{-2}

$$r \simeq 1 \times 10^{17} \ (n/10^9 \text{ cm}^{-3})^{-1/2} \text{ cm}, \tag{9}$$

$$f \simeq 1 \times 10^{-3} \ (n10^9 \text{ cm}^{-3})^{-1/2} .$$

This method applies only if there is full spatial coverage of the X-ray source by the BLR. If the spatial coverage is large but not complete, the absorbing column density could be very small part of the time, and a time average of N_H should be considered. A statistical analysis of N_H as deduced from the X-ray absorption, for a given class of active nuclei, could provide a clue for the spatial coverage.

The emission line spectrum is similar for Seyfert I galaxies for a large range of luminosities. The density of the BLR must then be rather similar for different nuclei. The few observed X-ray absorbing column densities are of the same order (which may arise from an observational selection effect). For given values of n and N_H, the overall radius of the BLR (or f^{-1}) scales as the square root of the luminosity of the Balmer lines. For large luminosities, f becomes very small and one may then expect that the spatial coverage decreases with increasing luminosity. A comparison between N_H obtained for nearby active galaxies and distant quasars with the forthcoming HEAO-B X-ray satellite may solve this problem.

If the observed motions within the BLR result from gravitational attraction of the central object, one can estimate the mass and Schwarzschild radius of this latter. One gets for NGC 4151

$$M = rv_{obs}^2 / G = 2.7 \times 10^8 \ (n/10^9 \text{ cm}^{-3})^{-1/2} \ M_o , \tag{10}$$

$$r_s = 2r \ (v_{obs}/c)^2 = 8 \times 10^{13} \ (n/10^9 \text{ cm}^{-3})^{-1/2} \text{ cm}.$$

b. Free-Free Absorption at Radio Wavelengths

The gas of the BLR is partly or fully ionized as could be inferred from the optical and UV line spectrum. The large column density derived from the X-ray absorption is a measure of the electron column density.

These electrons will greatly absorb the radio photons. The free-free optical depth at centimeter wavelengths λ is given by

$$\tau_{ff}(\lambda) = 2 \times 10^{-29}\, n_e\, N_H\, (T/10^4\ K)^{-3/2}\, \lambda^2 . \tag{11}$$

For $N_H = 3 \times 10^{22}\ cm^{-2}$ and $n_e \geq 10^8\ cm^{-3}$ one gets $\tau_{ff}(\lambda) > 60\, \lambda^2$.

A straightforward implication is that the BLR must not surround the radio source. The resolved radio source size is of the order of 1 pc for the BLRG 3C 120 and 3C 390.3 (Kellermann 1978). The size of the BLR is larger than that deduced for NGC 4151 but still smaller than 1 pc for densities larger than $10^8\ cm^{-3}$.

c. Surface Brightness of the Emission Line

As emphasized by M. Rees, the surface brightness in the core of an emission line cannot exceed that of a black body. This gives a lower limit on the size of the BLR

$$r \geq r\,(BB) = 3 \times 10^{14}\, (L/10^{42}\ erg\ sec^{-1})^{1/2}\, (T/10^4\ K)^{-2}\ cm . \tag{12}$$

This limit on r provides an upper limit of $nf^{1/2}$. For NGC 4151 one obtains $n\, f^{1/2} \leq 2 \times 10^{11}\ cm^{-3}$. As we must have $n \leq 10^{11}\ cm^{-3}$ and $f \ll 1$ (variability of N_H), one concludes than r should be much larger than r_{BB} for NGC 4151.

d. Optical and UV Absorption Lines

Absorption lines of HI and HeI (Anderson 1974) and CIV (Boksenberg *et al.* 1978) have been observed for NGC 4151. The true depth of these lines is not well determined but it largely exceeds the height of the continuum for both Hα and CIV. The absorbing region is thus outside the BLR, a conclusion reached by the authors mentioned above.

No intrinsic absorption from the BLR itself is then detected. We have evaluated the column density of HI from photoionization models and found that the optical depth in the Balmer lines is larger than 10 for NGC 4151, for a velocity dispersion in the line of sight of a few 10 km sec^{-1} (about 1 percent of the velocity inferred from the emission line width). The absence of HI absorption, in the case of a small velocity dispersion, is a consequence of the geometrical configuration of the BLR. Detailed transfer models are required to get information about the geometrical configuration of both the BLR and the UV continuum source.

e. Line Correlations: Is the BLR Composed of Distinct Emissive Regions?

There is no clear correlation between the optical line intensity of different lines. In particular the equivalent width, EW, of lines of high (HeII) and low (FeII) excitation do not show any trend of anti-correlation. One possible reason is the large scatter due to different excitation conditions (e.g. relative importance of electron collisions). An alternative explanation would be to assume the existence of different regions within the BLR: the FeII line region would be mechanically heated (Collin-Souffrin 1978) and the Balmer line region would be dominated by radiative input of energy.

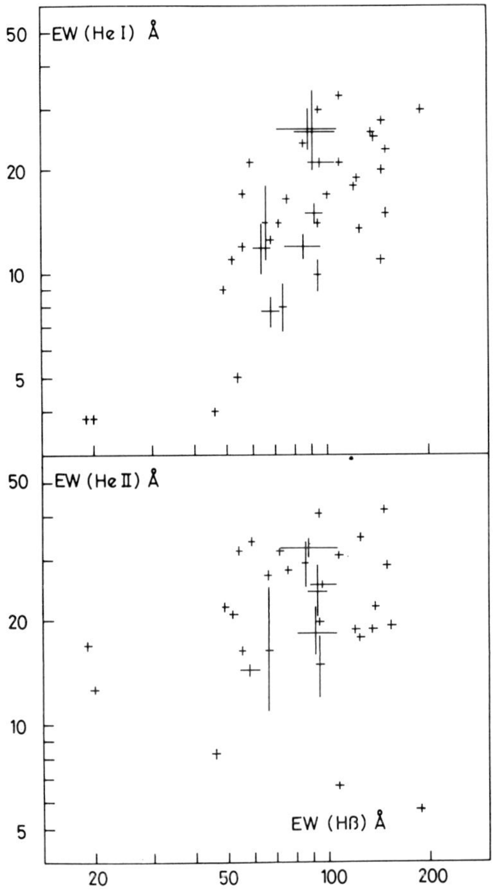

Fig. 3. Equivalent widths of HeIλ5876 and HeIIλ4686 as a function of the equivalent width of Hβ. The bars represent the differences between results given by different observers.

A weak correlation between the EW of HI and HeI lines may be present as shown in Figure 3. It could be compared with the lack of relationship between the EW of HI and HeII lines. The HI and HeI lines are most probably emitted in the same region, photoionized and of large optical depth.

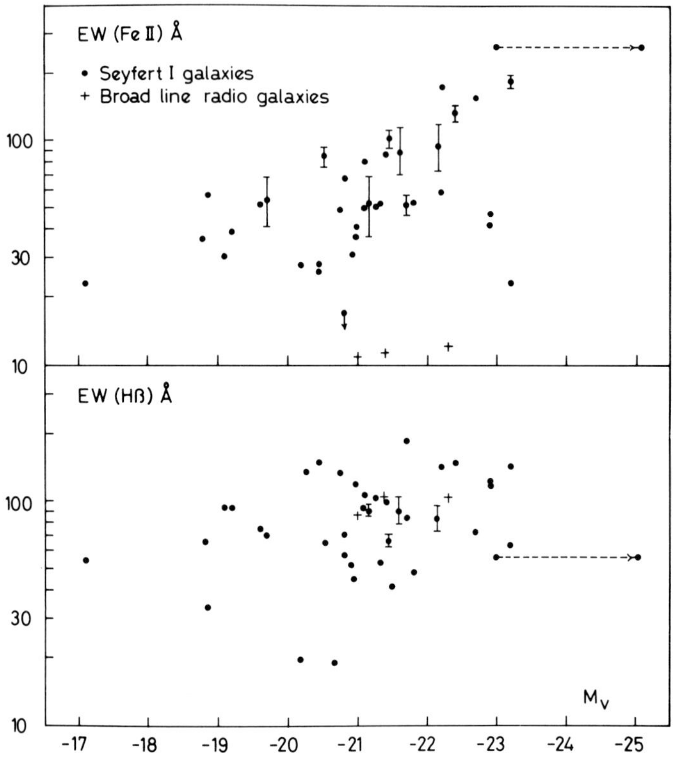

Fig. 4. Equivalent widths of FeIIλ4570, 5190, 5320 and Hβ as a function of the absolute optical magnitude. The vertical bars represent the differences between results given for the EW by different authors. The horizontal dashs account for a reddening correction.

The EW of Hβ is independent of the continuum luminosity, M_V, and is in the range 20 - 200 Å (as first noticed by Searle and Sargent 1968), but, as can be seen from Figure 4, the EW of FeII increases with M_V (as also noticed by Kunth and Sargent 1978). The range of values for EW(Hβ) is that expected from optically thick photoionized regions. However contrary to the suggestion of Searle and Sargent (1968) there is no trend of decrease of EW(Hβ) or EW(FeII) with increasing slope of the continuum ($I_\nu \propto \nu^{-\alpha}$, with α in the optical range as given by Stein and Weedman 1976 and Philipps 1978). In the frame of optically thick photoionized gas this may be accounted for by either 1) an absence of correlation between the optical and Lyman continua or 2) a large optical depth, and similar energy content in the very hard ionizing photons, as suggested by the relative X-ray power $L_x/L_{opt} \simeq 1$.

The correlation EW(FeII), M_V is inverse to that found by Baldwin (1977) between EW (CIV) and the UV luminosity for high redshift quasars (relationship tighter for radio flat quasars). The EW(FeII) shows an upper limit roughly following EW_{max} (FeII)$\propto L^{0.4}$. The spread of EW(FeII) and the existence of an upper limit may reflect the variation of the spatial coverage of the FeII region for different nuclei. The increase of EW(FeII) with M_V could be explained by an increase of energy input in the FeII region with increasing M_V.

The EW are proportional to the spatial coverage, $\Omega/4\pi$, of the corresponding emitting region. For Hβ we cannot argue, as was suggested for FeII, that $\Omega/4\pi$ greatly varies or must be small. If forthcoming X-ray observations confirm the existence of a low energy X-ray turnover for most Seyfert galaxies, this will confirm that $\Omega/4\pi$ of the BLR is always large for these galaxies. They must then have a strong Lyman discontinuity, in contrast to the high redshift quasars.

REFERENCES

Anderson, K.S.: 1974, Astrophys. J. 189, 195.
Auriemma, G., Angeloni, L., Belli, B.M., Bernardi, A., Cardini, D., Costa, E., Emanuele, A., Giovannelli, F., and Ubertini, P.: 1978, Astrophys. J. 221, L7.
Baldwin, J.A.: 1977, Astrophys. J. 214, 679.
Barr, P., White, N.E., Sanford, P.W., and Ives, J.C.: 1977, Mon. Not. R. Astron. Soc. 181, 43P.
Boksenberg, A., Snijders, M.A.J., Wilson, R., Benvenuti, P., Clavell, J., Macchetto, F., Penston, M., Boggess, A., Gull, T.R., Gondhalekar, P., Lane, A.L., Turnrose, B., Wu, C.C., Burton, W.M., Smith, A., Bertola, F., Capaccioli, M., Elvius, A.M., Fosbury, R.A.E., Tarenghi, M., Ulrich, M.H., Hackney, R.L., Jordan, C., Perola, C.G., Roeder, R.C., Schmidt, M.: 1978, Nature 275, 404.
Boldt, E.A.: 1978, XXIst COSPAR plenary meeting, Innsbruck, Austria.
Collin-Souffrin, S.: 1978, Phys. Scr. 17, 293.
Elvis, M.: 1976, Mon. Not. R. Astron. Soc. 177, 7P.
Elvis, M., Maccacaro, T., Wilson, A.S., Ward, M.I., Penston, M.V., Fosbury, R.A.E., and Perola, G.C.: 1978, Mon. Not. R. Astron. Soc. 183, 129.
Giacconi, R.: 1978, Phys. Scr. 17, 159.
Kellermann, K.I.: 1978, Phys. Scr. 17, 257.
Khachikian, E.YE., and Weedman, D.W.: 1974, Astrophys. J. 192, 581.
Kunth, D., and Sargent, W.L.W.: 1978, preprint.
Lyutyi, V.M.: 1973, Soviet Astron. A.J. 16, 763.
Mushotzky, R.F., Serlemitsos, P.J., Becker, R.H., Boldt, E.A., and Holt, S.S.: 1978, Astrophys. J. 220, 790.
Neugebauer, G.: 1978, Phys. Scr. 17, 149.
Neugebauer, G., Becklin, E.E., Oke, J.B., and Searle, L.: 1976, Astrophys. J. 205, 29.
Osterbrock, D.E.: 1971, *Nuclei of Galaxies*, p. 151, Pontificia Academia Scientiarum, ed. D.J.K. O'Connell.
Osterbrock, D.E.: 1978, Phys. Scr. 17, 285.
Penston, M.V., Penston, M.J., Selmes, R.A., Becklin, E.E., and

Neugebauer, G.: 1974, Mon. Not. R. Astron. Soc. 169, 357.
Philipps, M.M.: 1978, Lick Obs. Bull. No 802.
Pounds, K.: 1978, XXIst COSPAR plenary meeting, Innsbruck, Austria.
Rieke, G.H.: 1978, preprint.
Rieke, G.H., and Low, F.J.: 1975, Astrophys. J. 200, L67.
Schönfelder, V.: 1978, Nature 274, 344.
Searle, L., and Sargent, W.L.W.: 1968, Astrophys. J. 153, 1003.
Stark, J.P., Bell Burnell, J., and Culhane, J.L.: 1978, Mon. Not. R. Astron. Soc. 182, 23P.
Stein, W.A., and Weedman, D.W.: 1976, Astrophys. J. 205, 44.
Swanenburg, B.N., Bennett, K., Bignami, G.F., Caraveo, P., Hermsen, W. Kanbach, G., Masnou, J.L., Mayer-Hasselwander, H.A., Paul, J.A., Sacco, B., Scarsi, L., and Wills, L.D.: 1978, Nature 275, 298.
Tananbaum, H., Peters, G., Forman. W., Giacconi, R., and Jones. C.: 1978, Astrophys. J. 223, 74.
Telesco, C.M. Harper, D.A., and Loewenstein, R.F.: 1976, Astrophys. J. 203, L53.
Ward, J., Wilson, A.S., Disney, M.J., Elvis, M. and Maccacaro, T.: 1977, Astron. Astrophys. 59, L19.
Weedman, D.W.: 1977, Annu. Rev. Astron. Astrophys. 15, 69.
Wilson, A.S., Penston, M.V., Fosbury, R.A.E., and Boksenberg, A.: 1976, Mon. Not. R. Astron. Soc. 177, 673.

THE DENSITY-WAVE THEORY OF SPIRAL GALAXIES - COMPARISON BETWEEN THEORY AND OBSERVATIONS

Roland Wielen
Astronomisches Rechen-Institut, Heidelberg, Germany

Density waves are probably the most general phenomenon producing spiral structure. Density-wave theory is able to give a fairly successful interpretation of the observed spiral structure and of the related kinematics in external galaxies and in our Galaxy.

For a meaningful comparison between observations and density-wave theory, the proper theoretical devices should be used. Density-wave theory in the common version makes predictions only for objects in a stationary state: the linear theory should be used for older stars only; for HI, the strongly non-linear shock waves are appropriate in most galaxies with well-developed spiral structure. Young stars and short-lived objects, such as HII regions, are dynamically not in a stationary state and cannot be described by simple formulae. The drift of ageing spiral arms relative to the shock front is neither linear nor monotonic with age. The kinematical behaviour of young objects is also rather complicated. Furthermore, the behaviour of young objects depends sensitively on some additional assumptions about star formation, especially on the initial velocities with which the stars are born (e.g., pre- or post-shock velocities). All this hampers severely any conclusive comparison between theory and observations for younger stars and HII regions.

External galaxies now provide the best evidence for the existence of density waves in general. Good examples are M81 for the structure and kinematics of HI arms, and M51 for a shock front. The well observed galaxy M33 has probably only a weak density wave without a significant shock front, and is therefore less suited for a confrontation with theory.

The most convincing evidence for a density wave in our Galaxy is still the wavy irregularities in the apparent rotation curve as derived from the extreme radial velocities of HI at different longitudes. Most of the other observations in our Galaxy can be well explained by the presence of a density wave, but are less suited as decisive tests at present.

Bengt E. Westerlund (ed.), Stars and Star Systems, 81–83.

While the empirical data strongly suggest the existence of density waves in spiral galaxies, the observations have not provided any significant clue for the origin and maintenance of such density waves.

Details and specific references are given in a paper by the author which will be published in the proceedings of the IAU Symposium No. 84 (see general references). At the IAU Symposium No. 84, Vera Rubin presented interesting new results on the streaming motions of HII regions in the spiral arms of the Sc-galaxy NGC 2998 (V. Rubin, W.K. Ford, N. Thonnard: 1978, preprint, to appear in Astrophys.J.).

GENERAL REFERENCES

IAU Symposium No.38 : *The Spiral Structure of our Galaxy*, Eds. W. Becker and G. Contopoulos, Reidel Publ. Co., Dordrecht, 1970.
IAU Symposium No.58 : *The Formation and Dynamics of Galaxies*, Ed. J.R. Shakeshaft, Reidel Publ. Co., Dordrecht, 1974.
IAU Symposium No.60 : *Galactic Radio Astronomy*, Eds. F.J. Kerr and S.C. Simonson, Reidel Publ. Co., Dordrecht, 1974.
CNRS Colloquium No.241 : *La dynamique des galaxies spirales*, Ed. L. Weliachew, Editions du CNRS, Paris, 1975.
IAU Symposium No.77 : *Structure and Properties of Nearby Galaxies*, Eds. E.M. Berkhuijsen and R. Wielebinski, Reidel Publ. Co., Dordrecht, 1978.
IAU Symposium No.84 : *The Large-Scale Characteristics of the Galaxy*, Ed. W.B. Burton, Reidel Publ. Co., Dordrecht (to be published).

SOME ADDITIONAL REVIEW PAPERS

Burton, W.B.: 1973, Publ.Astron.Soc.Pacific 85, 679.
Burton, W.B.: 1976, Ann.Rev.Astron.Astrophys. 14, 275.
Kalnajs, A.J.: 1973, Proc.Astron.Soc.Australia 2, 174.
Lin, C.C.: 1975, in *Structure and Evolution of Galaxies*, Ed. G. Setti, NATO ASI C21, Reidel Publ. Co., Dordrecht, p. 119.
Lin, C.C., Shu, F.H.: 1970, in *Galactic Astronomy* Vol. 2, Eds, H.-Y. Chiu and A. Muriel, Gordon and Breach Sci.Publ., New York, p.1.
Marochnik, L.S., Suchkov, A.A.: 1974, Usp.Fiz.Nauk 112, 275 = Sov.Phys.Usp. 17, 85 (English translation).
Roberts, W.W.: 1977, Vistas in Astronomy 19, 91.
Roberts, W.W.: 1977, in *The Structure and Content of the Galaxy and Galactic Gamma Rays*, Eds. C.E. Fichtel and F.W. Stecker, NASA CP-002, Washington, p.119.
Rohlfs, K.: 1977, *Lectures on Density Wave Theory*, Lecture Notes in Physics Vol. 69, Eds. J. Ehlers *et al.*, Springer-Verlag, Berlin.
Rohlfs, K.: 1978, Mitt.Astron.Ges. No.43 (in press).
Schmidt-Kaler, T.: 1977, Vistas in Astronomy 19, 69.
Shu, F.H.: 1973, American Scientist 61, 524.
Toomre, A.: 1977, Ann.Rev.Astron.Astrophys. 15, 437.

van der Kruit, P.C., Allen, R.J.: 1976, Ann.Rev.Astron.Astrophys. 14, 417.

Wielen, R.: 1974, Publ.Astron.Soc.Pacific 86, 341.

Wielen, R.: 1975, in *Conference on Optical Observing Programs on Galactic Structure and Dynamics*, Ed. T. Schmidt-Kaler, Bochum, p.59.

STAR FORMATION AND GALACTIC STRUCTURE

Gas Density, Molecular Content, and Star Formation Rate in Spiral and Irregular I Galaxies

J. Guibert
Observatoire de Meudon, France

ABSTRACT

We recall the results of previous works which support the idea that, in spiral and irregular I galaxies, the star formation rate is governed by a law of the type: $d\rho_*/dt \sim K (\rho_g)^n$ (Schmidt 1959), with $n \sim 2$, ρ_g and ρ_* being the volume density of gas and young objects, as seems established for our Galaxy. The mean rates of star formation in 15 galaxies of the Local Group (Lequeux 1978), are examined. Applicability of a universal Schmidt's law ($n \sim 2$, K depending only on the type of young objects considered) to spirals and irregulars, cannot be ruled out within the uncertainties, the agreement being more easily obtained if the relative amount of molecular gas in late type, hydrogen-rich galaxies is significantly lower than in the solar neighbourhood in accordance with negative detections of molecules. However a decrease of K with increasing M_{HI}/M_{tot} is suggested. Among the possible explanations are the absence of spiral compression and the deficiency in heavy elements. This factors, together with a lower volume gas density, could also be responsible for a slight enrichment of the Initial Mass Function in massive stars with respect to the local IMF. A similar interpretation could account for the possible differences in star formation efficiency and Initial Mass Function between the solar neighbourhood and the galactic molecular 5 kpc ring.

"Let me say at first ... how struck I am by the delicate symbiosis that exists between the stars and the interstellar medium, how each is nourished by the other, and how the Galaxy as we know it is entirely a consequence of that balance and interplay"

G.H. Herbig, IAU symposium no 75 on star formation

"You crunch gas, and you get plenty of light"

A. Toomre (discussion following the paper by R.B. Larsson, Yale Conference, 1977).

Bengt E. Westerlund (ed.), Stars and Star Systems, 85–103.

The purpose of this paper is not to review the different topics related to star formation (S.F.) and galactic evolution, which have been treated in recent meetings (among them: Geneva, Yale, or Torun). I rather intend to examine a few points connecting the gas density to S.F. in spirals and irregular I galaxies (Early-type galaxies are treated by C.A. Norman in the present meeting). The main development will concern the relative importance of gas density, spiral structure, and heavy element abundance as regards star formation in 15 galaxies of the Local Group as compared to the solar neighbourhood. A few words will be said about the "intergalactic clouds". Finally, some problems concerning the central regions and the 5 kpc molecular ring of our Galaxy will be discussed.

I. STAR FORMATION RATE AND GAS DENSITY: THE SCHMIDT'S (1959) HYPOTHESIS

I.1. Schmidt's law.

According to Schmidt (1959), the star formation rate per unit volume could be proportional to a power n of the local gas density ρ_g. This implies that the volume density of young objects could be written

$$\rho_* = K(\rho_g)^n \qquad (1)$$

For the solar neighbourhood, Schmidt suggested $n \sim 2$.

Schmidt's law can be checked in two different ways. First, by searching for a correlation between ρ_* and ρ_g according to (1); second, by comparing the scale heights of gas and young objects. If the vertical gas distribution follows:

$$\rho_g = \rho_{g_o} \exp [- (z/a)^m] \qquad (2)$$

(m equal, e.g., 1 and 2 for exponential and gaussian distribution, respectively), and if (1) is assumed, then:

$$\rho_* = \rho_{*_o} \exp \{-(z/[a/n^{1/m}])^m\} \qquad (3)$$

ρ_{g_o} and ρ_{*_o} being the volume densities of the gas and the young objects near the galactic plane. Thus, if $n > 1$, the scale height z_* of the young population must be smaller than the scale height z_g of the gas by a factor of $n^{1/m}$:

$$z_* = z_g /(n^{1/m}) \qquad (3')$$

I.2. Schmidt's law in our Galaxy.

Preliminary results concerning the applicability of Schmidt's (1959) law relative to S.F. had been presented at IAU Colloquium number 45 (Guibert *et al.* 1977). From the HI and CO distributions by Baker and Burton (1975), and Burton and Gordon (1978), and using various assumptions for the ratio n(CO)/n(H2), Guibert *et al.* (1978) derived three models of total (molecular plus atomic) galactic hydrogen distribution. Using the most recent data concerning O and B stars and associations, cepheids, radio and optical HII regions, supernova remnants, pulsars and recombination lines, they have shown that the star formation rate (S.F.R.) in the Galaxy is consistent with Schmidt's law with:

$n \simeq 1.7 \pm 0.4$ (model 1) and
$n \simeq 2.1 \pm 0.5$ (model 2),

the vertical gas distribution corresponding to $m \sim 1.4$.

These results were obtained by combining the comparison of scale heights and the correlations of volume densities of gas and young objects (model 3, corresponding to a gradient of carbon abundance, gives a bad fit with Schmidt's law; see the paper for a discussion of the derivation of the CO and H_2 densities). It can be concluded that, under the assumption of a constant ratio n(CO)/n(H2) over the galactic disk, observations are consistent with a Schmidt's law with $n \simeq 1.9 \pm 0.6$, in agreement with Schmidt's (1959) estimates based only on HI data near the Sun.

I.3 Schmidt's law in external galaxies.

Apart from a few exceptions (see below I.3.2), the gas scale heights are unknown in external galaxies. Thus, in general, it is only possible to correlate surface densities σ_g and σ_* of gas and young objects. As will be seen in sect. I.3.1 the corresponding exponent, n', eventually found for surface densities: $\sigma_* = k\ (\sigma_g)^{n'}$, will differ from n as soon as the gas scale height varies over the galaxy. Moreover, though molecules have been detected in some galaxies (see II.4.5), the distribution of H_2 is unknown in these objects at the present time. Therefore this section is devoted to correlations of *atomic* neutral hydrogen with young objects.

I.3.1. Correlations between HI and young objects surface densities. From compilations by Martin *et al.* (1976), Berkhuijsen (1977) and Azzopardi and Vigneau (1977), it follows that, for M31, M33, NGC 2403, M 51, NGC 6946, M 101, NGC 6822 (Gottesman and Weliachew 1977a), the LMC and the SMC, n' is $\sim 2.0 \pm 0.6$. However (see e.g. Azzopardi and Vigneau 1977), correlations between σ_g and σ_* in inner and in outer parts of galaxies lead to discrepant results. On the average, $n'_{inner} \sim 1$, while $n'_{outer} \sim 2.5$. Moreover, for a given σ_{HI}, σ_* is higher in the inner regions. As noted by Madore *et al.* (1974), the latter phenomenon can be easily understood if the variation of z_g with the galactocentric distance r is taken into account. More precisely, for a given σ_{HI}, using (1) and (3') we get:

$$\sigma_*(\text{inner}) \,/\, \sigma_*(\text{outer}) = [z_g(\text{outer})/z_g(\text{inner})]^{n-1} \qquad (4)$$

therefore, as soon as $n > 1$, a larger thickness in the outer regions leads to a smaller apparent SFR as derived from surface densities. The interpretation of the apparent discrepancy between the slopes of the correlations ($n'_{inner} < n'_{outer}$) is less straightforward. However, if z_g is an increasing function of r, while HI presents a ring shape distribution (both conditions are frequently fulfilled), then, from:

$$n' = d\,Log(\sigma_*)\,/dLog\,(\sigma_g) = n - (n-1)\,dLog(z_g)/dLog(\sigma_g) \qquad (5)$$

it results that $n'_{inner} < n$, and $n'_{outer} > n$, in agreement with the observations, as soon as n exceeds unity. Note that, when the relative amount and scale height of molecules are unknown, only qualitative results can be derived from (4) and (5).

1.3.2. Determination of the gas scale height. For our Galaxy, the gas scale heights relative to cloud and intercloud HI, as well as to molecules, can be estimated by model fitting using observations at various latitudes (see e.g. Baker and Burton 1975; Burton and Gordon 1978). For external galaxies three ways of deriving z_g have been used up to now.

a/ Spheroidal shell model: A mass model comprising spheroidal shells of constant axial ratio q_o (cf. Burbidge *et al.* 1959) can be used to determine the total mass density ρ_o near the plane of the galactic disk by fitting a polynomial to the rotation curve. This method has been used by Warner *et al.* (1973) for M 33 (with $q_o = 0.2$)and by Emerson (1976) for M 31 (with $q_o = 0.1$). From ρ_o and the total width to half intensity of the velocity profiles, ΔV, the total gas thickness between half density points $2z_{1/2}$ can be derived (for a gaussian vertical distribution) by means of:

$$2\,z_{1/2} = \Delta V/(4\pi G\,\rho_o)^{1/2} \qquad (6) \qquad \text{or}$$

$$2\,z_{1/2}\,(pc) = 4.3\,\Delta V\,(km\,s^{-1})\,[\rho_o\,(M_\odot pc^{-3})]^{-1/2} \qquad (7)$$

(see, e.g. Talbot and Arnett 1975). The gas volume density is hence:

$$\rho g_o\,(M_\odot pc^{-3}) \sim \sigma_g\,(M_\odot pc^{-2})\,[\rho_o(M_\odot pc^{-3})]^{1/2}/\,\{4.6\,\Delta V(km\,s^{-1})\} \qquad (8)$$

The axial ratio q_o can be estimated from the inclination i and the ellipticity (b/a) of the isophotes. Its determination is highly sensitive to uncertainties affecting these two quantities:

$$(q_o)^2 = [(b/a)^2 - \sin^2 i]\,/\,[1 - \sin^2 i] \qquad (9)$$

A similar method has been used by Monnet (1971) for M 33.

b/ Density waves theory: The dispersion relation of the linear theory of Density Waves extended by Shu (1970) to disks of stars and gas of finite thickness involves the scale height z_g. Though the behaviour of the gas in spirals should rather be described by the non-linear theory (e.g. Roberts 1969; see also the paper by Wielen, this meeting), the linear theory can be used to investigate the behaviour of z_g as a function of r. By solving a set of equations including the dispersion relation proposed by Shu (1970) or the treatment of resonances by Lynden-Bell and Kalnajs (1972) Viallefond (1976) has obtained estimates of the gas scale heights for M 31 and M 33. The derived z_g, though smaller than those provided by method a) also indicate an increase of z_g with r (Fig. 1).

c/ HI observations of edge-on spirals: High resolution observations of NGC 891 have been performed by Sancisi and Allen (1978). From the isophotal contours in the plane (radial velocity; distance to the plane) they have shown that z_g is an increasing function of r.

The gas scale heights obtained so far for M 31, M 33, and NGC 891 appear in Fig. 1 together with the scale heights in our Galaxy (Baker and Burton 1975 for HI clouds; Guibert *et al.* 1978, for HI + H2). The scale height given by Warner *et al.* (1973) for M 33 has been divided by $\sqrt{2}$ to account for (6). For the purpose of comparison, using Schmidt (1956) mass density model for our Galaxy and after scaling to put the Sun at r = 10 kpc, we have drawn in Fig. 1 the scale height derived from (6) for the HI cloud medium, assuming a constant velocity dispersion $\sigma_{vz} = 6.4$ km s^{-1} taken from Falgarone and Lequeux (1973) for the solar neighbourhood. It can be noted that the scale heights obtained for NGC 891 (Sancisi and Allen 1978), for our Galaxy (Baker and Burton 1975; Guibert *et al.* 1978) as well as, from the density wave theory, for M 31 and M 33 (Viallefond 1976), increase faster than r in the outer regions in agreement with the thickness derived from Schmidt's (1956) model. The method of spheroids yields a linear increase of z_g (Warner *et al.* 1973; Emerson 1976) or a thickness which varies as $r^{1/2}$ (Monnet 1971). However, all the results indicate that z_g increases with r, which is easily predicted from (6), if the gas velocity dispersion is nearly constant with r (which seems actually to be the general situation) as soon as ρ_o decreases with increasing r.

I.3.3. Correlations between HI and young objects volume densities. Using the estimate of z_{HI} (sect. I.3.2) it is possible to plot ρ_* against ρ_{HI} in a log diagram. It can be seen in the case of M 33 (Fig. 2) that, when Viallefond's (1976) gas scale height is used to estimate the volume density, the discrepancy noted by Madore *et al.* (1974) between the inner and outer regions ($n' \sim 0.75$ for r < 2.4 kpc; $n' \sim 2.5$ for r > 2.4 kpc), disappears, leading to a unique correlation (with $n \sim 2.2$ for the optical and radio data used here). A similar improvement in the correlation has been obtained in the case of M 31 by Berkhuijsen (1977) using the scale height by Emerson (1976): n is found to be $\sim$ 1.8. For the same galaxy (and different observational data), Viallefond's (1976) thickness leads to $n \sim 1.5$. Note that Hamajima and Toja (1975) found $n'_{inner} \sim 1.1$ and $n'_{outer} \sim 3.1$. The important result to emphasize here is the reconciliation between inner and outer parts for M 31 and M 33 when *volume densities* are correlated.

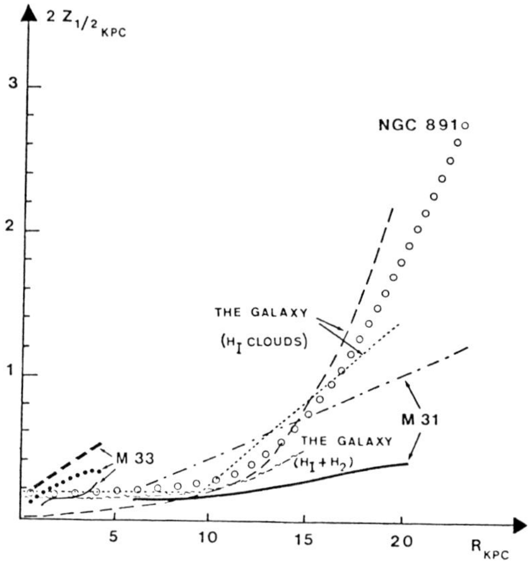

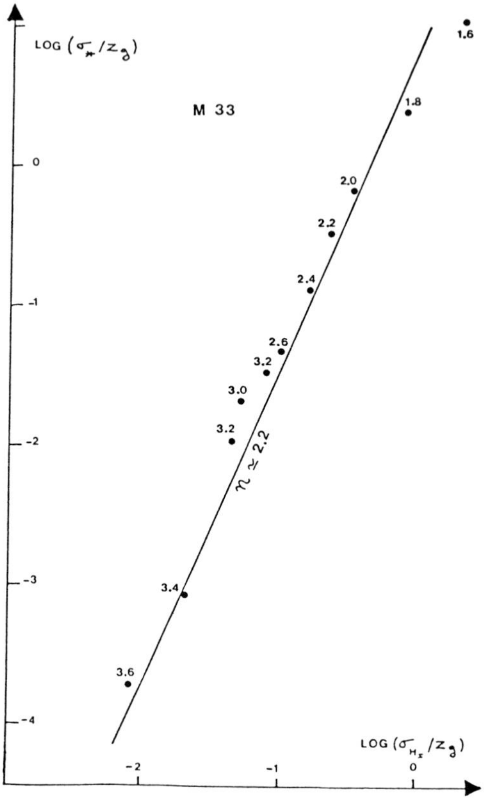

Figure 1. Gas scale heights in four spiral galaxies.

The Galaxy

HI Clouds: a/ dotted line: from Baker and Burton (1975)

b/ dashed line: from Schmidt's (1956) model

HI + H2: From Guibert *et al.* (1978)

M 31

dot-dashed line: Emerson (1976)
full line: Viallefond (1976)

M 33

dotted line: Monnet (1971)
dashed line: Warner *et al.* 1973)
full line: Viallefond (1976)

NGC 891

circles: Sancisi and Allen (1978)

Figure 2. Correlation between volume densities of gas and young stars in M 33. σ_{HI}: Gordon (1971)
σ_* : de Vaucouleurs (1968)
z_g : Viallefond (1976)

Comparison of ρ_* and ρ_{HI} in the Galaxy would contradict Schmidt's assumption, since the surface density of HI alone does not show the marked maximum near r $\sim$ 5 kpc as do the molecules and the young objects. It is rather surprising that in M 31 and M 33 ρ_* seems to be correlated with HI alone. This could indicate that, in these galaxies, molecules either are less abundant than in the Galaxy or follow a distribution similar to that of HI.

For M 31 and M 33 the value found for n depends not only on the observational data and on the method used to determine the scale heights but also on smoothing effects which can alter the true value of the slope. Moreover, as pointed out by Talbot (1971), detailed calculations involving assumptions about the initial mass function (I.M.F.) are required when different types of objects (e.g. bright stars, HII regions ...) are considered.

Conclusion. Keeping in mind all the above restrictions we can suggest that for our Galaxy, for M 31 and for M 33, the correlations between neutral gas and young objects volume densities seem to support Schmidt's assumption with $n \sim 1.8 \pm 0.5$. The similitudes observed in the correlations between surface densities suggest that this result could apply to all spiral and irregular I galaxies.

II. RATES OF STAR FORMATION IN SPIRAL AND IRREGULAR GALAXIES

In section I we have dealt only with the exponent of ρ_g in (1); n seems to be $\sim$ 2 in the galaxies which have been considered. Is the "constant" K (for a given type of young object) galaxy independent? Lequeux (1978) has compared the rates of formation of massive stars derived from counts in identical parts of the upper HR diagram for the solar neighbourhood, M 31, M33, and 4 magellanic irregular galaxies (NGC 6822, IC 1613, the LMC and the SMC). Except in M 31 (for which no star counts have been published) the SFR per unit gas mass (as derived from N_* / M_{HI}) seems to be a decreasing function of the ratio M_{HI} / M_{total} (and hence of the morphological type). Comparison of the number of massive stars and of other tracers of S.F. does not suggest important differences between the IMFs of the galaxies considered. This led us to consider the logarithmic mean of the various estimates of S.F. appearing in Table 1 and Fig. 1 of Lequeux's paper. This allowed us to compare the values of the SFR for the complete sample of 15 galaxies, from M 31 to the extremely hydrogen rich dwarf irregulars. The resulting estimate of the SFR per unit HI mass normalized to the solar neighbourhood, $E1 = (N*/M_{HI})/(N*/M_{HI})_\odot$, decreases in a smoother, but similar way to that suggested by Lequeux's Fig. 1. The galaxies are now closer to the curve temptatively drawn by Lequeux to fit his results at least for $M_{HI}/M_{tot} \lesssim 0.2$, particularly for M 31 which was the most puzzling case (see Table 1).

We now attempt to interpret the decrease of E1 with increasing M_{HI}/M_{tot} (and hence with increasing morphological type) in terms of

Schmidt's law.

II.1 Basic relation.

The results of section I suggest the investigation of the possible influence of ρ_g on the values found for El. Indeed, if $n \sim 2$, it is clear that:

$$NM = N_*/M_{gas} \propto \int \rho_g^{\,2}\, dv \,/ \int \rho_g\, dv = \int \rho_g\, dm \,/ \int dm = \langle \rho_g \rangle \qquad (10)$$

Therefore, NM turns out to be the average value $\langle \rho_g \rangle$ of ρ_g weighted by the mass of gas over the galaxy considered, and, as suggested by Lequeux (1978), a low value of $E = NM / NM_\odot$ may result from a low value of $R = \langle \rho \rangle_g / \rho_{g\odot}$.

II.2 Variation of the mean HI volume density along the Hubble sequence.

Since $\rho_g = \rho_{HI} + \rho_{H2}$ cannot be estimated directly we first investigate the possible variation of $\langle \sigma_{HI} \rangle$ with type. This quantity is expected to decrease with increasing type T from Sa to Irr I galaxies on the following grounds.

First, there is a general (weak) decrease of the "true HI surface density" as defined by Bottinelli (1971); the factor is ~ 2 from Sb to Irr I systems. However, the scatter is important.

Second, if the gas velocity dispersion is assumed not to decrease, the gas scale height is expected to increase according to (6). This is due to the decrease of the total mass density $\bar{\rho}_o$. Such a decrease was first suggested by Holmberg (1964). Even if we correct by a factor of 3 - for Sc and Irr I - the total masses used by Holmberg, following the suggestion by Sandage *et al.* (1970), using a mean true axial ratio $q_o = 0.5$ for Irr I (de Vaucouleurs and Pence 1973), it turns out that $\bar{\rho}_o$ decreases by a factor of ~ 3 between Sa and Sb-Sc systems, and by ~ 5 between S_b and Irr I. Note that Rubin *et al.* (1978) find that the total surface density decreases by a factor of 4 from Sa to Sc galaxies. Inference from this result as regards the volume density would require taking account of the evolution of the diameter and axial ratio. Bottinelli and Gougenheim (1974) questioned the decrease of $\bar{\rho}_o$ with type. However, within a given type their estimate of $\bar{\rho}_o$ decreases with decreasing B-V colour in agreement with the results from Holmberg and Rubin *et al.*

We can conclude that from Sb to Irr I the HI volume density ρ_{HI} is expected to decrease according to (8) by a factor of about $2\sqrt{5} \sim 5$ with a noticeable dispersion.

Note, by the way, that if the total mass density, observed at the present time, reflects the initial gas density of the protogalaxy, then the initial gas density was probably higher in early than in late type galaxies, which could - at least partly - account for the smaller remaining relative amount of gas observed at the present time in early systems, due to the high efficiency of S.F. according to (10). Such a suggestion was made by van den Bergh (1959) and Holmberg (1964). However, the rough consistency of the increase in M_{HI}/M_{tot} with the decrease in $\overline{\rho}_o$ between S_b and Irr I galaxies is to handle with caution. According to Larson (1976 a, b), to obtain a significant disc component with protogalactic collapse models it is necessary to assume a greater drop in the star formation rate during the later stages of the collapse than is given by a power-law function of the gas density. The roles of angular momentum, tidal forces (Larson, 1977 a and b) ... are important. Moreover, relating the present total mass density to the initial gas density would require a good knowledge of the evolution of the velocity dispersion during galaxy formation and evolution.

II.3. Mean gas density and efficiency of star formation.

II.3.1. Mean HI density. Since H_2 is unknown, we first estimate $<\rho_{HI}>$ in the 15 galaxies of Lequeux's (1978) sample. For M 31 we use the HI surface density by Guibert (1974) and the gas scale height by Emerson (1976). For M 33, the HI surface density by Huchtmeier (1973) together with the gas scale height derived from Warner *et al.*'s (1973) data (see sect. I.3.2) were employed. For the remaining 13 galaxies we used the data referenced by Lequeux and estimated $<\rho_{HI}>$ by means of (8), assuming a full width ΔV of 28 km s^{-1} (see, e.g. Emerson 1976). The mean total mass density $\overline{\rho}_o$ was estimated from the total mass M_{tot} up to radius a_o and an assumed axial ratio q_o = 0.5 (de Vaucouleurs and Pence 1973): $\overline{\rho}_o \sim M_{tot}/(\frac{4}{3}\pi\ q_o\ a_o^3)$.

II.3.2. Check of the Schmidt's law. If the SFR is entirely governed by Schmidt's law (1) and if the relative amount of molecules is close to the local value, according to (10) the ratio

$$k1 = R1 / E1 \left[R1 = <\rho_{HI}> / \rho_{HI\odot};\ E1 = (N_* / M_{HI})/(N_* / M_{HI})_\odot \right]$$

is expected to be close to unity (assuming n = 2 and K equal for all systems).

If the ratio $<\rho_{H2}> / <\rho_{HI}>$ is x($x_\odot$ for the solar neighbourhood) and assuming that the scale heights for HI and H2 are roughly in a ratio of 2 throughout the galaxy (this is the case in our Galaxy at least up to 10 kpc) the ratio: $M_g / M_{HI} = 1 + M_{H2} / M_{HI}$ is $\sim 1 + x/2$. Taking account of molecules leads us to consider instead of k1, the quantity k:

$$k = (<\rho_g> / \rho_{g\odot}) \Big/ \left\{ (N_*/M_g)/(N_*/M_g)_\odot \right\} = k1 \frac{(1+x)\ (1+x/2)}{(1 + x_\odot)(1 + x_\odot/2)} \qquad (11)$$

The quantity $(1+x_\odot)\ (1 + x_\odot/2)$ takes the value 3.5 and 2 in models 1 and 2 respectively.

Assuming Schmidt's law (thus k = 1) allows us to estimate x for a given galaxy. The limiting case (x = 0, no molecules) corresponds to $kl = (1 + x_\odot)\ (1 + x_\odot/2) = kl_{max}$.

Table 1. Star formation rate per unit mass of atomic hydrogen in 15 galaxies of the Local Group

	Morphological type	M_{HI}/M_{tot}	$R1 = \langle\rho_{HI}\rangle/\rho_{HI\odot}$	$E1 = \frac{(N_*/M_{HI})}{(N_*/M_{HI})_\odot}$	$k1 = R1 / E1$
M31	Sb	0.02	0.80	0.65	1.2
Solar neighbourhood		0.04	1	1	1
M33	Scd	0.05	0.6	0.6	1
II ZW 70	C	0.07	0.5	0.5	1
NGC 6822	IBM	0.07	1.	0.60	1.7
LMC	SBm	0.09	1	0.8	1.3
IC 1613	Im	0.15	0.3	0.3	1
IC 10	IBm	0.19	1.4	0.2	7
DDO 125	Im	0.21	0.4	0.3	1.3
II ZW 71	C	0.21	1.7	0.7	2.4
II ZW 40	C	<0.44	>0.5	0.2	> 2.5
NGC 3109	Sm/Im	0.22	0.24	0.05	4.8
SMC	IBm	0.32	1.7	0.3	5.7
HoI	Im	0.32	0.74	0.1	7.4
Sag DIG	Im	0.38	0.27	0.04	7
S DIG	Im	0.36	0.65	0.035	20

References for the data can be found in the paper by Lequeux (1978) except for the mean HI volume densities which have been estimated in the present paper.

The results of our calculations are displayed in Table 1. The galaxies are listed in order of increasing M_{HI}/M_{tot}. Though there is a noticeable scatter in the evolution of the derived parameters (particularly as regards IC 10 and DDO 125), the first impression is that kl increases with M_{HI}/M_{tot}, at least for $M_{HI}/M_{tot} \gtrsim 0.2$. Beyond this value our estimate of kl exceeds 3.5 which suggests that in hydrogen rich galaxies, if the SFR is governed by a law of Schmidt's type, the constant K of eq. (1) is smaller than for galaxies of earlier type, indicating that SF proceeds less efficiently.

However, the (large) uncertainties must be taken into account. Errors can reach a factor of 2 both in N_*/M_{HI} and in M_{HI}/M_{tot}. Uncertainties of at least a factor of 2 probably affect σ_{HI}. Errors on M_{tot} affect the gas scale height. As a result the estimate of kl may be in error by a factor of 5, particularly for dwarf, faint galaxies for which sensitivity and resolution problems can be important.

Consequently, it is not firmly established that star formation proceeds less efficiently in late type, gas rich systems and equation (1) could be valid everywhere with the same K and n. Agreement with a universal Schmidt's law would be more easily obtained if the molecular content of extreme irregulars is negligible and if the solar neighbourhood is relatively molecular-rich (model 1 of Guibert *et al.* 1978).

If the trend suggested by Table 1 is real, the sample galaxies appear to be roughly divided into two classes, the transition occurring for $M_{HI}/M_{tot} \sim 0.2$.

a/ For seven galaxies: M 31, M 33, the LMC, NGC 6822, IC 1613, II Zw 70 and DDO 125 (hereafter class I), kl is found to lie between 0.5 and 2. Within the uncertainties indicated by Lequeux (1978) this means that the mean SFR in these galaxies seems to be consistent with a *unique* Schmidt's law with $n \sim 2$ and that the relative amount of molecules is not significantly different from the local value [$\sim$ 40 % and $\sim$ 25 % for Guibert *et al.*'s (1978) models 1 and 2, respectively].

b/ For six galaxies: NGC 3109, IC10, the SMC. Ho I, Sag DIG and S DIG (hereafter class II), kl exceeds 5. Even if molecules are absent (which favours SF near the Sun) the SFR in these galaxies is lower, when compared to class I, than could be predicted from a universal Schmidt's law.

For instance for the group NGC 3109, Ho I, Sag DIG and S DIG, which present the weakest SFR per unit gas mass, $R1 = \langle \rho_{HI} \rangle / \rho_{HI\odot} \simeq 0.5$, $E1 = (N_*/M_{HI})/N_*/M_{HI})_\odot$ is ~ 0.05; the ratio $kl = R1/E1$ is of the order of 10 and the low gas density fails by a factor of 3 - 5 (and even more for S DIG) to account for the low SFR.

c/ In two intermediate cases (II Zw 40 and II Zw 71) $kl \sim 2.5$, suggesting a low molecular content and a marginal consistency with Schmidt's law, assuming model 1 for the solar neighbourhood (10 %, instead of 40 % as near the Sun, of the mass of the gas are then in molecular form in these two compact galaxies). It could be tempting to explain the relatively high SFR of these two systems by compactness or interaction-induced burst of S.F. However, DDO 125, with a relative hydrogen content (0.2) similar to that of II Zw 71, seems to show a similar S.F. efficiency relative to its density (kl = 1.3) and is neither compact nor interacting. On the other hand, IC 10 appears to be abnormally inactive for its relative hydrogen content.

II.4 Interpretation.

The above results concern a limited sample of galaxies. The uncertainties are large. It is not firmly established that SF proceeds less efficiently in late type systems and applicability of a universal Schmidt's law ($n \sim 2$, K depending only on the type of young objects considered) cannot be ruled out. However, it is interesting to investigate the possible mechanisms which could be responsible for the decrease of the "constant" K suggested by our calculations. Among the possible interpretations we shall consider: the inhibition of S.F. by tidal forces, the role of a possible threshold, the influence of bursts of star formation, the role of density waves, and the role of heavy elements in the process of star formation.

II.4.1. Tidal forces. Free fall collapse of clouds caused by self-gravity would lead to $n \sim 1.5$ (Madore 1977). Larson (1976a) suggested that collapse may be inhibited by tidal forces:

$$d\rho_* / dt = A \rho_g^{\,n} \left[1 + B (\bar{\rho}_{tot} / \rho_g)\right]^{-1} \qquad (B \sim 1) \qquad (12)$$

$\bar{\rho}_{tot}$ corresponding to the mean density of matter interior to the region considered. It is not clear whether such effects could be more efficient in late type than in early type galaxies, particularly in view of the high relative gas content of irregulars.

II.4.2. Gas density threshold for star formation. Goldreich and Lynden-Bell (1965) studied the growth of gravitational instabilities in a differentially rotating disk and derived a critical density depending on Oort constants. The value suggested for the solar region (2-3 cm^{-3}) is higher than the adopted density (~ 1 cm^{-3}), which could indicate that some additional compression mechanisms could be needed.

II.4.3 Bursts of star formation. Bursts of star formation are apparently occuring in the LMC, as suggested by cepheids (Payne- Gaposchkin 1971), clusters (Hodge 1973), and supergiants (Ardeberg 1975). Such bursts could be operating in the interacting pair II Zw 70-71 (Balkowski *et al.* 1978, O'Connell *et al.* 1978), and in II Zw 40 (Gottesman and Weliachew 1977 b). However, it seems unlikely that such phenomena could explain the difference suggested between our two classes of nearby galaxies as regards the SFR.

II.4.4. Compression by density waves. Roberts *et al.* (1975) plotted the velocity of the gas perpendicular to the spiral arms against van den Bergh's (1960) luminosity class. The frequency F of encounters between the gas and the spiral wave appears to be correlated with the intensity of the spiral structure and to decrease with increasing type T. It is interesting to note that NGC 3109 (magellanic spiral or irregular ?) is located at the lower end of the diagram, which suggests that a low SFR could be due to the weakness of the spiral wave compression. However the same diagram suggests a significantly higher SFR for M 31 than for M33, which is not obvious in the data compiled by Lequeux (1978).

According to Jensen *et al.* (1976) the abundance of heavy elements in galactic disks is correlated with the frequency F, which is a decreasing function of T and r. However, even when no spiral structure is assumed, models of galactic evolution, involving an initial gas surface density decreasing with increasing r, predict a heavy-element abundance which decreases outwards. This prediction is in agreement with current ideas concerning the abundance of heavy elements in spirals (e.g. Peimbert 1975). However a recent work by Collin-Souffrin *et al.* (1978) reveals through a reinterpretation of spectroscopic observations that a clear trend of increasing heavy element abundance towards the centre only appears in M 101 and NGC 2403. This result, which is not inconsistent with a real gradient in the solar neighbourhood (see e.g. Mayor 1976) could be explained by a permanent large scale mixing of the gas by non-circular motions in the plane of the disks or by infall or accretion of intergalactic matter. Despite the above objections it is doubtless that compression induced by spiral waves can play a significant role in the S.F. process.

II.4.5 Roles of heavy elements in the star formation process. According to the results of sect. II.3 it cannot be excluded that (at least in galaxies with M_{HI}/M_{tot} in the range 0.1 - 0.2 and perhaps for higher relative gas contents) S.F. continues to be governed by a Schmidt's law with n and K similar to those relevant to galaxies of earlier types. This agreement with a universal Schmidt's law would be more easily obtained if the proportion of molecules in late type galaxies is significantly lower than in the solar neighbourhood. Such a low (or possibly negligible) molecular abundance would be in agreement with the fact that, up to now, CO has been unsuccessfully searched for in 22 irregular I galaxies. Molecules (CO, OH, H2CO, HCN, H2O) have been detected only in 11 Sb-Scd galaxies, in the Irr II galaxy M 82, and in the LMC which is a barred spiral with $M_{HI}/M_{tot} \lesssim 0.1$ (see e.g. Combes *et al.* 1978). For galaxies with detected CO emission no correlation is found between CO emission and HI content. These results are quite consistent with the irregulars being relatively unevolved systems, still rich in atomic gas, in which the abundance of heavy elements (responsible, for instance, for the formation H_2 onto dust grains) is very low, and seems to decrease with increasing T. For instance, for young stars of Pop. I, Z $\sim$ 0.03 in the Galaxy, 0.025 in the LMC and 0.01 in the SMC where $M_{HI}/M_{tot} \sim 0.3$ (van den Bergh 1975). With respect to the solar neighbourhood O/H $\sim$ 0.8 in M 33, 0.56 in the LMC, 0.51 in NGC 6822, and 0.19 in the SMC(for references see Lequeux 1978). The gas to dust ratio is $\sim$ 9 in the LMC and $\gtrsim$ 15 in the SMC as compared to the local value (Azzopardi and Vigneau 1977).

A significant deficiency in heavy elements in late type galaxies could perhaps explain the low SFR per unit mass. Talbot and Arnett (1973) and Talbot (1974) have developed models of star formation in which the abundance Z of heavy elements plays an important role. The critical pressure to produce the thermal instability (and hence, presumably, through collapse, star formation) in the interstellar gas is roughly inversely proportional to Z. Therefore, the probability for SF to occur can be very low in metal poor regions according to these models of metal-enhanced star formation (MESF).

Conclusion: The apparently low SFR per unit mass of gas in the most hydrogen-rich galaxies of the local group studied by Lequeux (1978) could be explained by: a low volume gas density and/or the absence of spiral waves and/or a deficiency in heavy elements.

II.5. The Initial Mass Function .

Examination of the number of giant and supergiant H II regions per unit mass of gas and its comparison with N_*/M_{HI} suggests that the upper IMF might be richer in very massive stars in the Magellanic Clouds and in M 33 than in the Galaxy. (Lequeux 1978). These - slight - differences in the IMF with respect to the solar neighbourhood could have the same origin(s) as the apparently low SFR in late type systems.

II.5.1 Role of the density. The proportion of very massive stars is expected to increase in regions of low density. This is a consequence of the expression giving the minimum unstable mass M_J (Jeans' mass) for which the collapse of a cloud of gas density ρ_c can occur: (Jeans 1928):

$$M_J = 1.86\ (RT/G)^{3/2}\ \rho_c^{-1/2} \tag{13}$$

Burki (1977) has shown that the initital upper mass spectrum is steeper in the case of small galactic clusters which are preferentially formed at smaller distances from the galactic centre where the gas density is higher. He has also established that the upper part of the IMF is steeper in the central, dense regions of the newly formed clusters and that many very massive stars are formed in the outer regions. (Burki 1978).

However, our estimates of $<\rho_g> / \rho_{g\odot}$ for M 33, the LMC and the SMC are ∿ 0.57, 0.85 and 0.77, respectively, and do not differ significantly from unity.

II.5.2. Gas compression by spiral shocks. The increase in the linear radius of young clusters with the distance to the galactic centre, a consequence of (13), has been interpreted by Burki and Maeder (1977) as due partly to a decrease in the mean gas density $\overline{\rho}_g$, partly to a decrease in the effective gas compression (EGC) (e.g. Roberts and Yuan 1970), from 8 to 12 kpc from the centre, the gas density in the cloud being the product of ρ_g by EGC. The decrease of the shock strength with increasing type could enrich the upper region of the IMF.

II.5.3. Role of Z. The Jeans' mass is expected to be greater if interstellar clouds are cooled less efficiently when depleted in heavy elements and the upper mass limit for SF may be greater in very metal-poor conditions (e.g. Larson and Starrfield 1971; Kahn 1974). However, according to Larson (1977a) spiral structure could under certain conditions induce formation of very massive stars. The role of the density (II.5.1) is also controverted. A high number of - *non visible* - M dwarfs in the - low density - outer regions of spirals has been invoked (e.g. Roberts and Whitehurst 1975) to account for the flat rotation curves. However, the ages of these stars are unknown (this would require spectroscopic observations or kinematical studies) and it is far from certain that they are at the present time being formed from gas of so low density. At r $\gtrsim$ 25 kpc in M 31, for instance, ρ_{HI} is $\sim$ 100 times smaller than in the central regions and the SFR is probably negligible. By the way, it seems possible to give a negative answer to the question asked by King (1977): will the optical size of galactic disks grow in the future?

II.6. Tentative conclusion.

The applicability of a universal Schmidt's law to irregulars as well as spirals cannot be ruled out at least in the Local Group. Such a law would of course concern only the SFR averaged over each galaxy and would not exclude local variation due to peculiar conditions. If the decrease in the SFR per unit mass of gas suggested by Lequeux (1978) for hydrogen rich galaxies is real, it can be explained by a relatively low HI volume density, a small molecular content, the absence of gas compression by spiral shocks, or the underabundance in heavy elements.

Most of these factors could also be responsible for slight differences in the IMF, some galaxies appearing to have a proportion of very massive stars higher than the solar neighbourhood. Improvement of the observational data concerning the galaxies of the Local Group and more accurate determination of the tridimensional gas distribution would be of considerable help in the development of theories of star formation.

III. ABSENCE OF OPTICAL COUNTERPART TO "INTERGALACTIC CLOUDS"

HI clouds without optical counterparts have been observed in the direction of groups or pairs of galaxies (Mathewson *et al.* 1975; Cesarsky *et al.* 1977; Balkowski *et al.* 1978). We do not intend to discuss here the problem of the distance of such clouds (are they parts of extragalactic groups, or of the Magellanic Stream, as suggested by Haynes and Roberts, 1978?). Our purpose is to show that the absence of optical counterparts can be easily understood in terms of Schmidt's law. For a system with a single fluid the plane parallel solution for the volume density is (e.g. Ledoux 1951):

$$\rho_o(M_\odot pc^{-3})^{1/2} = 0.15\ N_H(10^{20} cm^{-2}) / \Delta V\ (km\ s^{-1}) \qquad (14)$$

N_H being the projected hydrogen column density. It is important to note that this estimate of ρ_o is distance independent. For the "intergalactic cloud" observed by Cesarsky *et al.* (1978), assuming a full width ΔV = 28 km s^{-1} (as found in M 31 by Emerson 1976), we obtain:

$$\rho_o \sim 10^{-5}\ M_\odot pc^{-3}$$

(Dividing the HI mass by the volume of the sphere corresponding to the diameter indicated by Cesarsky *et al.* (1978) leads to a few 10^{-5} $M_\odot pc^{-3}$). This is at least 2 orders of magnitude lower than the lowest density found in section II (for comparison, near the Sun

$\rho_g = 1\text{-}2\ 10^{-2}\ M_\odot\ pc^{-3}$, depending on the abundance of CO). Therefore, the star formation, if any, should proceed at a negligible rate in these clouds. We have reached a similar conclusion in section II.5.3 for the HI extensions of galactic disks.

IV. THE GALACTIC CENTER AND THE 5 KPC RING

We now consider some consequences of S.F. on the structure of gas disks. We shall not discuss the models proposed by Mueller and Arnett (1976) and Gerola and Seiden (1978), according to which selfpropagating star formation (SPSF) can induce the formation of a spiral structure, because these models do not account for the spiral structure of the gas and are therefore beside our present preoccupations. Leaving our neighbours of the Local Group and the intergalactic clouds we return to our Galaxy and to some problems concerning the rate of gas consumption in the center and in the so called 5 kpc ring:

IV.1. What is the exact importance of the central gas deficiency?

CO observations by Gordon and Burton (1976) indicate a deep depression in the gas density inside r = 4 kpc. The surface density seems to be at least 4 times lower than in the region of the ring, the mass required to fill the hole being $7\ 10^8\ M_\odot$. However Sanders (1977) notes that a recent survey of the central regions by Bania (1977) suggests a mean gaseous density in the inner 600 pc in excess of $\sim 500\ M_\odot pc^{-2}$, which corresponds to $\sim 5\ 10^8\ M_\odot$. This suggests that inside 4 kpc the gas has experienced some efficient loss of angular momentum due either to viscous effects (Ostriker 1977) or perhaps to breaking by a rotating oval or bar-like distorsion (Sanders 1977). It is important to recall that estimates of the distribution of H_2 are affected by many uncertainties (see e.g. the discussion of this point by Guibert *et al.* 1978). Accurate estimate of the amount of gas consumed in the S.F. process in the central regions is not available at the present time.

IV.2. What is the origin of the 5 kpc ring?

A ring-shaped gas distribution can be easily explained by models of galactic evolution in which the SFR increases faster than the gas density (see e.g. the results obtained by Talbot and Arnett 1975, for n = 1.5 and 2). Gordon (1978), assuming that the inner Lindblad resonance occurs where the CO distribution exhibits a sharp discontinuity (at a radius of 4 kpc), determines a spiral pattern speed of 11.5 km s^{-1} kpc^{-1}, which is in agreement with previous estimates (Lin *et al.* 1969; see also Grosbøl 1977; and Palous 1977). In this picture the molecular ring corresponds to a maximum of gas compression leading to efficient cloud formation. Atomic hydrogen is indeed present at smaller values of r. However the coincidence between the maximum of total (HI + H_2) gas surface density with the inner Lindblad resonance would probably remain to be explained.

IV.3. Is the SFR consistent with a Schmidt's law in the 5 kpc ring?

The answer is positive as regards the data used by Guibert *et al.* (1978). Here we compare the gas density to the distribution:

1/ of the unidentified Type II OH/IR stars,

2/ of the stars responsible for the near and far infrared emission from the galactic plane.

IV.3.1. Unidentified Type II OH/IR stars. According to Bowers (1978) the surface density of these stars is higher at r $\sim$ 5 kpc than near the Sun by a factor of $\sim$ 10. This figure is higher than the corresponding ratio of the gas surface densities by factors of 3.7 and 5 (for models 1 and 2, respectively), which can be explained by a SFR governed by a Schmidt's law with n $\simeq$ 2 and 2.5, in agreement with the exponents mentioned in section I.2.

IV.3.2. Infrared data. From data by Puget *et al.* (1978), in the 5 kpc ring, the ratio total young Population I luminosity/ mass of gas, normalized to the solar neighbourhood, takes values in the ranges 7 - 11 and 10 - 15 for models 1 and 2, respectively. These figures exceed the factors of $\sim$ 3.7 and 5 quoted above by $\sim$ 2 - 3. It is not unreasonable to admit an overall uncertainty of that order; in such a case, all available data would fit with a SFR governed by the same dependence on the gas density (similar considerations would apply to the galactic center).

If we adopt the high rate of star formation suggested by Puget *et al.* (1978) the gas consumption time scale, which is $\sim 2\ 10^9$ yr near the Sun, is found to be $\sim 2\ 10^8$ yr in the 5 kpc ring, and the gas will be exhausted very quickly unless replenished by infall or some other mechanism. An alternative solution could be a steepening of the IMF with respect to that adopted for the solar vicinity, in agreement with the deficiency in HII regions in the inner galaxy deduced from the comparison of the Lyman continuum photons production with the total luminosity of young stars.

This suggested steepening of the IMF in dense regions is in agreement with the considerations developed in section II.5 about dependence of the critical masses on the mean gas density, the compression strength, and the abundance of heavy elements; these three parameters being likely higher in the ring than in the solar neighbourhood.

REFERENCES

IAU Symposium no. 75 *Star Formation.* Geneva 1976. Eds. T. de Jong and A. Maeder, D. Reidel, Dordrecht, 1977.

Yale University Conference. *The Evolution of Galaxies and Stellar Populations* 1977. Eds. B.M. Tinsley and R.B. Larson, Yale Univ. Obs., 1977.

IAU Colloquium no. 45. *Chemical and Dynamical Evolution of our Galaxy.* Torun, 1977. Eds. E. Basinska-Grzesik and M. Mayor, Geneva Obs. 1978.

Ardeberg, A.: 1975, Proc. 3d Europ. Astron. Meeting, Tbilissi, p. 193.
Azzopardi, M., Vigneau, J.: 1977, Astron. Astrophys. 56, 151.
Baker, P.L., Burton, W.B.: 1975, Astrophys. J. 198, 281.
Balkowski, C., Chamaraux, P., Weliachew, L.: 1978, Astron. Astrophys. in press.
Bania, T.M.: 1977, Astrophys. J. 216, 381.
Berkhnijsen, E.M.: 1977, Astron. Astrophys. 57, 9.
Bottinelli, L.: 1971, Astron. Astrophys. 10, 437.
Bowers, P.F.: 1978, Astron. Astrophys. 64, 307.
Burbidge, E.M., Burbidge, G.R., Prendergast, K.H.: 1959, Astrophys. J. 130, 739.
Burki, G.: 1977, Astron. Astrophys. 57, 135.
Burki, G.: 1978, Astron. Astrophys. 62, 159.
Burki, G., Maeder, A.: 1976, Astron. Astrophys. 51, 247.
Burton, W.B., Gordon, M.A.: 1978, Astron. Astrophys. 63, 7.
Cesarsky, D.A., Falgarone, E.G., Lequeux, J.: 1977, Astron. Astrophys. 59, L 5.
Collin-Souffrin, S., Joly, M., Vigroux, L.: 1978, in preparation.
Combes, F., Encrenaz, P., Lucas, R., Weliachew, W.: 1978, in preparation.
Emerson, D.T.: 1976, Mon. Not. R. Astron. Soc. 176, 321.
Falgarone, E., Lequeux, J.: 1973, Astron. Astrophys. 25, 253.
Gerola, H., Seiden, P.E.: 1978, Astrophys. J. 223, 129.
Goldreich, P., Lynden-Bell, D.: 1965, Mon. Not. R. Astron. Soc. 130, 125.
Gordon, K.J.: 1971, Astrophys. J. 169, 235.
Gordon. M.A.: 1978, Astrophys. J. 222, 100.
Gordon. M.A., Burton, W.A.: 1976, Astrophys. J. 208, 346.
Gottesman, S.T., Weliachew, W.: 1977, Astron. Astrophys. 61, 523.
Gottesman, S.T., Weliachew, L.: 1977 b, Astrophys. J. 211, 47.
Grosbøl, P.J.: 1977, IAU Colloq. no. 45, 279.
Guibert, J.: 1974, Astron. Astrophys. 30, 353.
Guibert, J., Lequeux, J., Viallefond, F.: 1977 IAU Colloq. no. 45, 165.
Guibert. J. Lequeux, J., Viallefond, F.: 1978, Astron. Astrophys. 68, 1.
Hamajima, K., Tosa, M.: 1975. Publ. Astron. Soc. Japan 27, 561.
Haynes, M.P., Roberts, M.S.: 1978, preprint.
Hodge, P.W.: 1973, Astron. J. 78, 807.
Holmberg, E.B.: 1964, Ark. Astron. 3, 387.
Huchtmeier, W.: 1973, Astron. Astrophys. 22, 91.
Jeans, J.H.: 1928, *Astronomy and Cosmogony,* Cambridge Univ. Press, Cambridge.
Jensen, E.B., Strom, K.M., Strom, S.E.: 1976, Astrophys. J. 209, 748.
Kahn, F.D.: 1974, Astron. Astrophys. 37, 149.
King, I.R.: 1977, Yale University Conference p. 1.
Larson. R.B.: 1976a, Mon. Not. R. Astron. Soc. 176, 31.
Larson. R.B.: 1976b, *Galaxies,* p. 67; Sixth Advanced Course of the Swiss Society of Astronomy and Astrophysics; Eds. L. Martinet and M. Mayor, Geneva Observatory.
Larson. R.B.: 1977a. Yale University Conference p. 97.
Larson, R.B.: 1977b. IAU Colloq. no. 45, p. 3.
Larson. R.B.: Starrfield, S., 1971, Astron. Astrophys. 13, 190.
Lequeux, J.: 1978, Astron. Astrophys. in press.
Lin, C.C., Yuan, C., Shu, F.H.: 1969, Astrophys. J. 155, 721.

Lynden-Bell, D., Kalnajs, A.J.: 1972, Mon. Not. R. Astron. Soc. 157, 1.
Madore, B.F.: 1977, Mon. Not. R. Astron. Soc. 178, 1.
Madore, B.F., van den Bergh, S., Rogstad, D.H.; 1974, Astrophys. J. 191, 317.
Martin, N., Prevot, L., Rebeirot, E., Rousseau, J.: 1976, Astron. Astrophys. 51, 31.
Mathewson, D.S., Cleary, M.N., Murray, J.D.: 1975, Astrophys. J. 195, L 97.
Mayor, M.: 1976, Astron. Astrophys. 48, 301.
Monnet, G.: 1971, Astron. Astrophys. 12, 379.
Mueller, M.W., Arnett, W.D.: 1976, Astrophys. J. 210, 670.
O'Connell, R.W., Thuan, T.X., Goldstein, S.J.: 1978, preprint.
Ostriker. J.P.: 1977, IAU Colloq. no. 45, p. 241.
Palous, J.: 1977, IAU Colloq. no. 45, p. 293.
Payne-Gaposchkin, C.H.: 1971, Smithsonian Contr. Astrophys. 13, 1.
Peimbert. M.: 1975, Annu. Rev. Astron. Astrophys. 13, 113.
Puget, J.L., Serra, G., Ryter, C.: 1978, in preparation.
Roberts, W.W.: 1969, Astrophys. J. 158, 123.
Roberts, W.W., Yuan, C.: 1970, Astrophys. J. 161, 877.
Roberts, W.W., Roberts, M.S., Shu, F.H.: 1975, Astrophys. J. 196, 381.
Roberts. M.S., Whitehurst, R.N.: 1975, Astrophys. J. 201, 327.
Rubin, V.C., Ford, W.K., Thonnard, N.: 1978, preprint.
Sancisi, R., Allen, R.J.: 1978, Astron. Astrophys. in press.
Sandage, A., Freeman, K.C., Stokes, N.R.: 1970, Astrophys. J. 160, 831.
Sanders, R.H.: 1977, IAU Colloq. no. 45, p. 103.
Schmidt, M.: 1956; Bull. Astron. Inst. Netherlands 13, 15.
Schmidt, M.: 1959; Astrophys. J. 129, 243.
Shu, F.H.: 1960, *Theory of Spiral Structure,* in *Galactic Astronomy;* Eds. H-Y Chiu and A. Muriel, Gordon and Breach.
Talbot, R.J.: 1971, Astrophys. Lett. 8, 111.
Talbot, R.J.: 1974, Astrophys. J. 189, 209.
Talbot, R.J., Arnett, W.D.: 1973, Astrophys. J. 186, 69.
Talbot, R.J., Arnett, W.D.: 1975, Astrophys. J., 197, 551.
Van den Bergh, S.: 1959, Publ. Astron. Soc. Pacific, 71, 5.
Van den Bergh, S.: 1960, Pub. D.D.O., II. no. 6.
Van den Bergh, S.: 1975, Annu. Rev. Astron. Astrophys. 13, 217.
Vaucouleurs, G. de: 1968, See ref. in Gordon, K.J. 1971: Astrophys. J. 169, 235.
Vaucouleurs, G. de, Pence, W.: 1973, Bull. American Astron. Soc. 5, 446.
Viallefond, F., 1976: Doctoral Thesis, Université Paris VII.
Warner, P.J., Wright, M.C.H., Baldwin, J.E.: 1973, Mon. Not. R. Astron. Soc. 163, 163.

RECENT RESULTS ON GALACTIC X-RAY SOURCES

J. Trümper
Max-Planck-Institut für Physik und Astrophysik, Institut für Extraterrestrische Physik, Garching, München.

One of the outstanding achievements of X-ray astronomy was the discovery of very bright compact objects in binary systems. The field has now reached a point where low luminosity X-ray sources like coronae of nearby stars have been detected. We briefly survey recent results obtained for normal stars, hot white dwarfs and U-Geminorum systems. The main part of this review will be devoted to accreting neutron stars in binary systems and X-ray burst sources. About 15 X-ray stars with regular pulsations have been found so far and most of them are undoubtedly accreting neutron stars. After surveying the general properties of these sources we shall discuss the complex variability and spectral behaviour of Her X-1 in some detail. In particular we refer to the 35 day cycle and the cyclotron lines found in the hard X-ray spectrum of this source which allow a spectroscopic determination of the magnetic field strength near the surface of the neutron star (Her X-1: 5×10^{12} Gauss). As far as bursters are concerned, more than 30 sources have been discovered so far. There is mounting evidence that most of the bursts are produced by accretion into compact objects of stellar mass. A particularly successful model assumes thermonuclear flashes on the surface of accreting neutron stars.

Bengt E. Westerlund (ed.), Stars and Star Systems, 105.

ON THE ACTIVITY OF THE NUCLEI OF GALAXIES

E. Ye. Khachikian
Byurakan Astrophysical Observatory, Armenia, USSR

INTRODUCTION

The history of science is aware of cases when the name of the author of one theory or concept or another, that has won universal acknowledgement, is often not mentioned or even forgotten. I do not believe that the name of the man who advanced and substantiated the idea of the activity of the nuclei of galaxies, Prof. Ambartsumian, is somehow forgotten. At present the interest in the problem has grown to such an extent and the papers on the activity of the nuclei of galaxies are so numerous that I think it proper to make a brief historical account on the subject of the discourse.

This fact achieves greater prominence today since this year is the 20th anniversary of the date when the first systematical statement of the concept of the activity of the nuclei of galaxies was first made public.

If some 30 or 35 years ago a discourse on a similar topic had been submitted it would have drawn at least an ironic smile. Indeed, the general view held some 30 years ago concerning the galaxies was the fact that they were thoroughly formed steady systems with a rich past and with no prospect of radical changes in the future. Therefore, the investigation of the structure of galaxies was in most cases confined to their classification and general photometry, based solely on their external morphological characteristics, setting little store by the composition of their central regions. Professor Ambartsumian was the first to pay attention to specific dynamic phenomena associated with the nuclei, particularly to the eruptive phenomena in them. The role of nuclei in the evolution of galaxies had manifestly been underestimated.

Since 1956 Ambartsumian has been advancing and developing the concept of the basic role of the nuclei of galaxies in their life and evolution (Ambartsumian 1958, 1962, 1965).

Bengt E. Westerlund (ed.), Stars and Star Systems, 107–122.

This idea was stimulated by a number of wonderful discoveries resulting in the revision of our notions on the world of galaxies. They are as follows:

a) The identification of one of the powerful radiosources Cygnus-A with the galaxy containing two nuclei (Baade and Minkowski 1954). A similar picture was observed also in the radiosource Perseus-A (NGC 1275).

b) The observation of ejections of conglomerates of clouds of relativistic electrons, of gas clouds and non-stable stars (NGC 4486) and blue condensations (NGC 3561a and IC 1182) from the nuclei of galaxies.

c) Outflow of matter from the centre of our Galaxy.

d) The discovery of Seyfert (Seyfert 1943) and Haro (Haro 1956) galaxies.

Here, Seyfert's paper, now regarded as classical, should be singled out for mention. The galaxies which he investigated (less than ten in number), are distinguished by the high luminosity of their nuclei and, more importantly, by the width of the Balmer emission lines. The great width of the emission lines indicates that the turbulent motions of gas clouds in the nuclei of those galaxies at times attain a velocity of over 3000 km/sec. Now it seems quite strange and surprising that this very important paper of Seyfert was not duly taken into account in the succeeding twenty years or so until the concept on the activity of the nuclei of galaxies was made public.

What is the implication of the "activity" of the nuclei of galaxies? According to Ambartsumian (see, for example Ambartsumian 1965), it manifests itself mainly in the following forms:

1. Outflow of ordinary gas matter (in form of jets or clouds) from the nuclear region at the velocity of up to hundreds of kilometres per second.

2. Continuous emission of the flux of relativistic particles or other agents, producing high energy particles, as a result of which a radio halo may form around the nucleus.

3. Eruptive ejections of gas matter (M82 type).

4. Eruptive ejections of concentrations of relativistic plasma (NGC 4486, 5128, etc.)

5. Ejection of compact blue condensations with an absolute magnitude of the order of the luminosity of dwarf galaxies (NGC 3561, IC 1182). In this case the division of the nucleus into two or more comparable components is also presumed, initiating the formation of multiple galaxies.

The presence of one or several of those phenomena allow us to call a nucleus active.

Now quite a few observational data are available and a number of interesting objects have been discovered that are in favour of the activity of galaxies, and supplement mentioned notions.

This is in the first place the discovery of quasars and optical quasars, new Seyfert galaxies and the detection of nonstable phenomena in their nuclei, BL Lac type objects, the observation of eruptive processes in the nuclei of galaxies and, finally, the discovery of a large number of galaxies with UV-excess.

Our aim is not to make a complete survey of those data; it is simply impossible to achieve this object in a single report or over a short period of time. Moreover, there is no need of it, as a collection of papers under the caption "Quasars and the active nuclei of galaxies" of the Copenhagen symposium saw print quite recently, in which all those problems have been elaborated.

We should like, however, by relying on the analysis of observational data of galaxies with UV-excess alone, to show how the activity is manifested in this or that form, in line with the concept of Ambartsumian.

GALAXIES WITH UV-EXCESS

The history of objects with UV-excess originated in the mid-sixties when, on Ambartsumian's initiative, Markarian began to make a successful survey of the sky with a view to detecting galaxies with anomalous spectra (Markarian 1967), in Byurakan, using a 40" telescope of the Schmidt system provided with an objective prism of the same diameter. The research is still underway, and over 1200 objects with an anomalous ultraviolet excess in their spectra as their characteristic feature have hitherto been found. To this number about 240 objects with the same characteristics should be added, detected by Kazarian, the list being prepared for print now.

Relying on the general view of the spectrum on plates taken with an objective prism, Markarian (1967) distinguishes two basic types of objects: "s" and "d". Galaxies attributed to the "s" type have narrow, continuous spectra sharply outlined all over the height, similar to those of stars. As to the nature of the energy distribution, those objects are similar to the quasi-stellar ones and to the nuclei of Seyfert galaxies. To the "d" type are ascribed the objects, the spectra of which have a diffuse form with a weak continuous spectrum. In general such spectra are similar to those of compact associations of blue stars and gas nebulae.

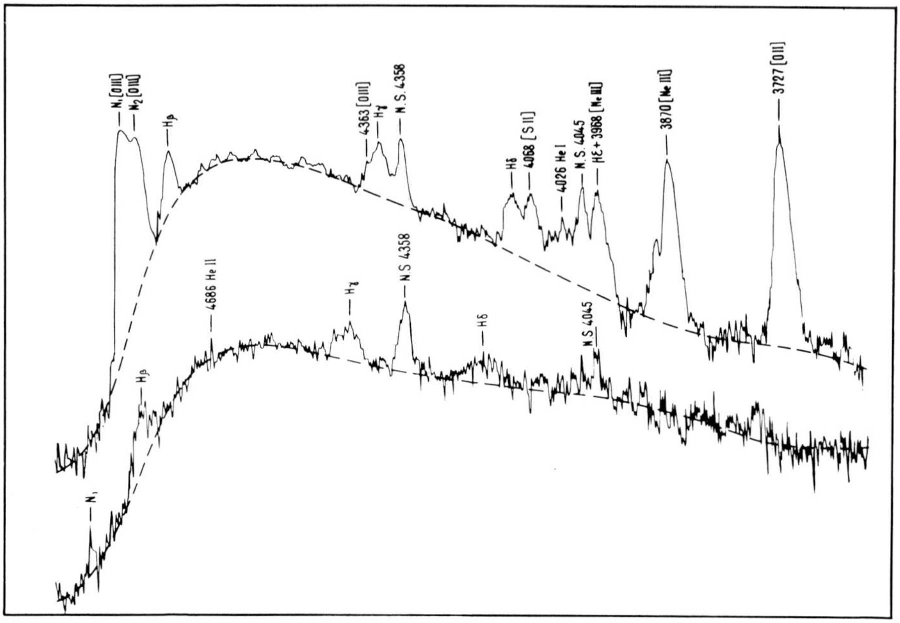

Fig. 1. Upper - spectrum tracing of Sy2 galaxy Mark. 3
lower - spectrum tracing of Sy1 galaxy Mark. 10.

The first detailed spectral investigations of galaxies with an ultraviolet excess indicated (Khachikian 1968, Arp, Khachikian, Linds, Weedman 1968; Weedman, Khachikian 1968, 1969) that they are quite interesting objects. Over 85 per cent of the Markarian objects turned out to own emission lines, their intensity being directly dependent on the value of the UV-excess, i.e. one can conclude that the *presence of a strong ultraviolet continuum is closely associated with the formation of the emission spectrum and the more intensive the continuous spectrum in the visible ultraviolet is, the more intensive are the emission lines.*

It also became evident that the spectra of those objects differ, nevertheless, essentially from each others as to the excitation degree of the emission lines and their widths (Khachikian 1968). Moreover, they turned out to differ sharply in morphological characteristics as well: one can come across the blue galaxies of Haro, the compact galaxies of Zwicky, the N type galaxies, the spiral and irregular galaxies among the Markarian objects. Quite important is the discovery of a number of Seyfert galaxies and quasars among those objects.

Another feature of the objects with the UV-continuum is the striking disparity of the values of red shifts, ranging over a great interval, between z= 0.002 and z = 1.75. That is why their luminosities differ greatly, as their visible magnitudes lie in the main in a comparatively narrow interval ($14\overset{m}{.}5$- $16\overset{m}{.}6$). In addition to supergiants with M_B =-22^m, objects are also met with the absolute magnitude in the order of - 13^m, i.e. in the order of luminosities of superassociations or bright associations (Ambartsumian 1964; Ambartsumian, Iskudarian, Shahbazian and Sahakian 1963). Markarian 132 with z = 1.75 is distinctive for its record absolute brightness: M_V = - 28^m (Sargent 1972). Thus even disregarding Markarian 132, which is an exclusive object, *the luminosities of objects with UV-continuum vary by four orders.*

It turns out that objects displaying one sign of activity or another, relating particularly to the galactic nucleus, are met with much more often among galaxies with a UV-continuum.

Let us consider some typical examples:

1. Seyfert Galaxies

About ten per cent of the Markarian galaxies show spectral signs typical of the Seyfert galaxies. We should like to dwell on some of those galaxies which spectroscopically have been studied in detail (Khachikian and Weedman 1971). As to the intensity of hydrogen emission and forbidden lines, the Seyfert galaxies turned out to be of two classes: Sy1 and Sy2. For a better illustration, Fig. 1 shows the registrograms of the spectra of Markarian 3 and 10 that represent those two classes of galaxies with wide lines. The spectra were obtained in the same night, on the same plate and with the same telescope (36" Crossley telescope with a nebular spectrograph.). Galaxies of the first type have very bright and wide hydrogen lines and quite narrow and faint forbidden lines [N II] , [O II] , [O III] , [Ne III] and so on. In them the ratio of the intensity of the lines of [O III] λλ 5007, 4959 to Hβ is less than 1. On the other hand, in galaxies of the second type, on the contrary, the forbidden lines are as wide as the hydrogen lines, whereas the intensities of the lines [O III] exceed by one order that for Hβ. Typical representatives of Sy1 are Markarian 9 and 10, the absolute brightness of their nuclei in U light is equal to $M_u = -21^m.3$ and $-20^m.4$, respectively. Among Sy1, Markarian 478 displays a record brightness with $M_u = -23^m.3$. The width of the hydrogen lines in those galaxies corresponds to Doppler velocities of the order of 3000-5000 km/sec. On the other hand, the nuclei of Sy1 are quite compact and star-like, similar to the quasars. Their bolometric luminosity is rather high - from 10^{42} to 10^{46} erg/sec (Stein and Weedman 1976).

Most of the known Seyfert galaxies relate to the Sy1 type.

The second type galaxies concede superiority to them with regard to the power of the ultraviolet excess and absolute brightness. However, the excitation and activity degree in Sy2 type galaxies is rather high. Markarian 6 provides a brilliant example. It was in Markarian 6 that the spectral changes indicating a possible explosion in the nucleus were first observed (Weedman and Khachikian 1970, 1971 b).

Spectroscopic observations of Markarian 6, made at intervals over a period of three years, show that within a year new emission components Hα and Hβ appeared, shifted from the basic line toward the short wavelengths to a value corresponding to the Doppler velocity of the order of 3000 km/sec. This is in favour of the appearance of a new gas cloud, of a very high radial velocity. The total mass of radiating matter proved to be, by rough estimates, of the order of one solar mass; while the mass of the whole hydrogen cloud ejected from the nucleus is estimated at several hundreds of solar masses (Gyulbudagian 1971).

Thus, *a flow of matter can take place from the nuclei of Sy2 galaxies, such a flow eventually influencing the outer structure of the galaxies as a whole.*

Further detailed spectrophotometric investigation of Markarian 6 led to the assumption that *not one but two hydrogen clouds apparently were ejected from the nucleus of this galaxy* (Khachikian 1973).

2. Markarian 7 and 8, NGC 6306

The morphological and spectral structure of the central regions of those galaxies with UV-excess unites them [the UV-excess in NGC 6306 has been detected by M.A. Kazarian (see Kazarian and Khachikian

1974)]. True, so far no thorough spectral investigation of Markarian 7 has been conducted, but its external similarity with Markarian 8 is so striking (Kalloglian 1971) that their spectral similarity is highly probable.

These galaxies are distinguished by a very interesting and unusual structure and dynamic peculiarities. Their central regions contain five bright condensations included in the diffuse envelope (Khachikian 1972; Kazarian and Khachikian 1977). The spectra of the condensations are similar and consist of bright emission lines of hydrogen and forbidden lines [S II]λλ 6717/31, N[II]λλ 6583/48, [O III]λλ 5007, 4959, [O II]λ 3727. The condensations are super-associations or, as they are frequently termed in literature, giant HII regions. For instance, the brightest condensation in Markarian 8 has $M_{pg} = -18^m$ (Khachikian 1972). Particular mention should be made of the difference in the radial velocities of individual condensations in the same galaxy, attaining as much as 600 km/sec (Kazarian and Khachikian 1977; Khachikian 1972). Fig.2 shows the picture of NGC 6306 taken with the 6 m telescope of the Special astrophysical observatory (SAO) of the USSR Academy of Sciences. The electron density in some condensations equals about $2 \times 10^3 cm^{-3}$ (Kazarian and Khachikian 1977).

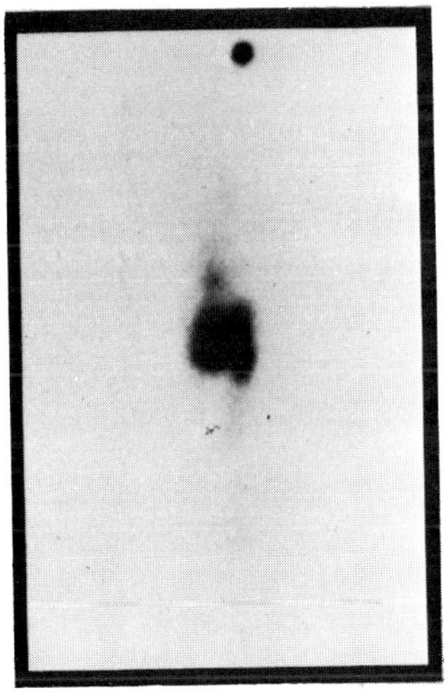

Fig. 2 A photo of NGC 6306. Prime focus of 6 m telescope of SAO AN USSR. (Kodak 103a - 0, Exp - 5 min., N - to the top, E - (to the left).

In those objects *the nuclei are seemingly split into separate condensations reminiscent of superassociations.*

3. Objects - "twins"-near Markarian 261 and 262

These blue objects are located at a distance of about 2' to the north of Markarian 261 and 262. In blue light they appear in the form of two compact condensations joined together by means of bars. In the red light they nearly disappear and the bars become barely visible. The distance between the objects - "twins"- is about 4 kps in the projection on the celestial sphere. Together with Markarian 261 and 262 - they have also similar red shifts - they form, no doubt, one physical system (Arp, Heidmann and Khachikian 1974).

A notable peculiarity of this system is, in the first place, the great similarity of the objects "twins", the appearance, dimensions, radial velocities, luminosities and spectra of which are nearly identical. Secondly, the lines joining the pair Markarian 261-262 and the objects "twins" are parallel, though the fortuity of coincidence is quite unlikely. It is interesting to note that there is also another pair of blue objects, again with a parallel direction to the former two, between those two pairs. Their radial velocities have not been measured but if this pair also turns out to possess the same red shift, thereby the uniqueness of the entire system will further be emphasized. This is probably the case of activity when the original body divides into separate components. Furthermore, there is good reason to presume that the system originated as a result of a successive fragmentation (Ambartsumian 1958). In other words, first the body from which the system of "twins" originated, separated from the body out of which Markarian 261-262 arose, followed by a division within each of those systems.

4. Galaxies with UV-excess with Multiple Nuclei

It became clear recently that objects with double and multiple nuclei (Petrossian, Sahakian and Khachikian 1978) are met with among galaxies with a UV-excess. Of the 620 galaxies investigated 59 possess double and multiple nuclei. Most of the components are of a size of the order of 1 kpc, and the double ones range from 1 to 3 kpc. The mutual distance of the components of multiple nuclei does not exceed 2 kpc and in those of double nuclei not 7 kpc.The absolute luminosity of multiple nuclei vary in the range of -9^m to -18^m, and that of double ones from -12^m to -21^m. Another important feature of multiple and double nuclei is the fact that an increase of the distance between the components of the nuclei brings about an intensification of their brightness. In addition, with the increase of this distance between the components of the nuclei new structural particulars appear more and more frequently (filaments, ejections, spiral arms in an embryonic and undeveloped state). Interestingly, Seyfert galaxies are also met with among UV galaxies with double nuclei (Markarian 463, 673, 739, 789). In general, objects with double nuclei occur more frequently among galaxies with a UV-excess than among other types of galaxies.

Fig. 3 gives the pictures of some galaxies with UV-excess with double nuclei.

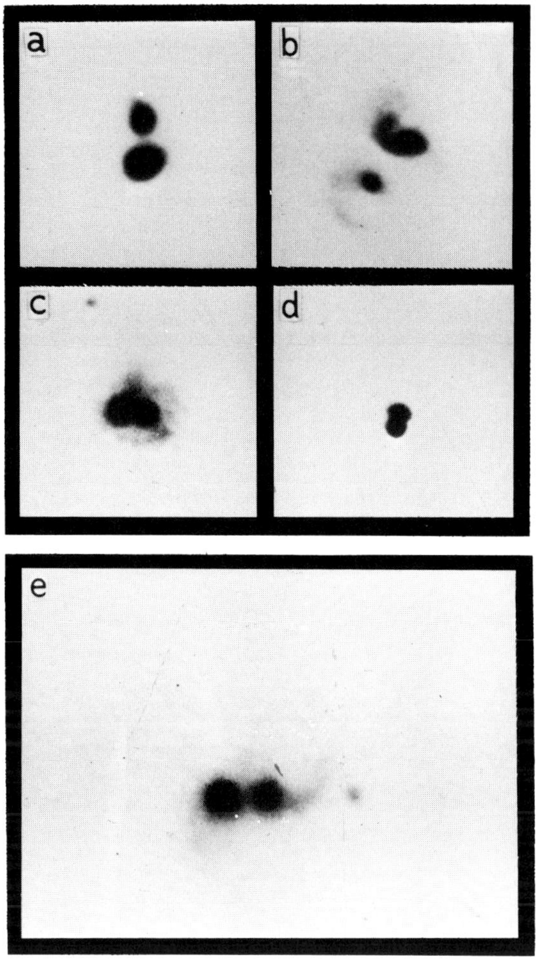

Fig. 3. Photos of double - nucleus Markarian galaxies in blue.
a. 480 (6 m telescope, exp. 5 min)
b. 551 (2.6 m telescope of Byurakan observatory, exp. 10 min.)
c. 739 (2.6 m telescope of Byurakan observatory, exp. 10 min.)
d. 930 (2.6 m telescope of Byurakan observatory, exp. 10 min.)
e. 212 (2.6 m telescope of Byurakan observatory, exp. 20 min.)
(Scale:1 mm ∿ 2").

These facts uphold the hypothesis of Ambartsumian on the division of the initial dense nucleus into separate components, their further withdrawal from each others and the formation of separate structural features.

a

Fig. 4 a. A photo of double-nucleus galaxy Mark. 266. (2.6 m telescope of Byurakan Observatory. ORWO-Zu-2, exp. 25 min.) Scale: 1 mm ∿ 1".5.

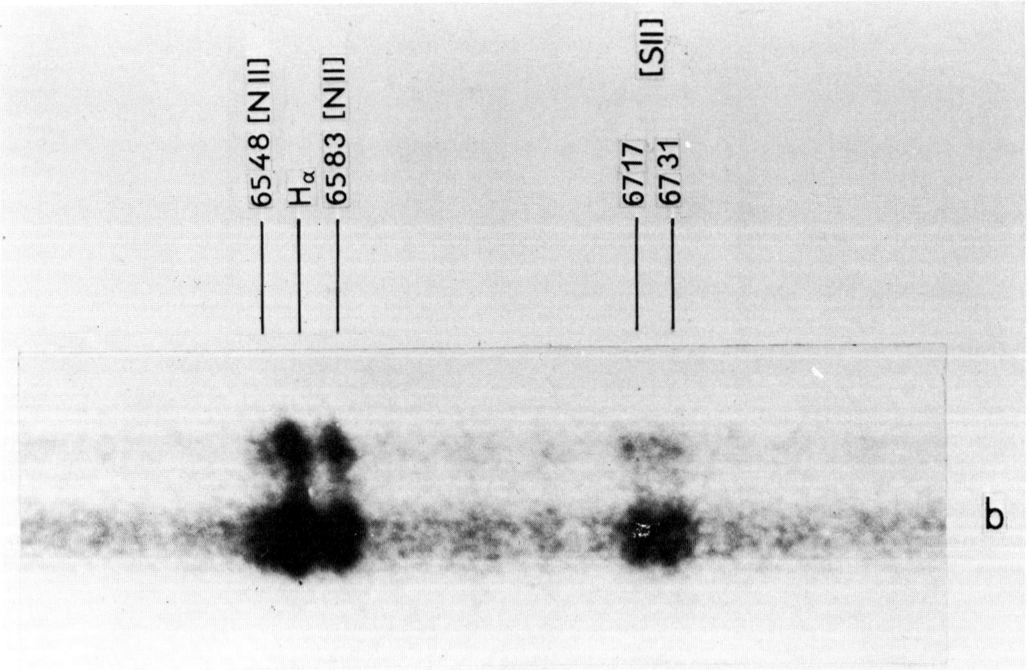

Fig. 4 b. Reproduction of a spectrum of Mark. 266 (6 m. telescope prime-focus spectrograph + IT. original dispersion ∿ 85 Å/mm.)

Recently, spectral observations of several galaxies with double nuclei have been made in the prime focus of 6m telescope of the SAO of the USSR Academy of Sciences on the spectrograph SP 160 with IT. (The research has been conducted by A.R. Petrossian, K.A. Sahakian and E.Ye. Khachikian, and a paper on the spectral investigations of double nuclei will soon be presented for publication). The results are rather interesting.

The observations show that the components of double nuclei are not always spectroscopically identical and sometimes differ markedly both in the form of the spectrum and in the intensity of the continuum and the lines. We shall not enlarge on the results of this research here, but with the permission of the co-authors, we shall quote data concerning one object, namely Markarian 266 (NGC 5256).

Markarian 266 is a galaxy with two very bright condensations in the centre (Fig. 4a). The northern condensation is more compact and star-like than the southern one. One spiral arm originates from each condensation - long and developed in the southern and tight and twisted in the northern. The absolute brightness of the components is correspondingly, equal to Mpg = - $17^{m}.8$ and Mpg = $-17^{m}.5$ and their dimensions are about 2.2 kpc, the distance between them equals 6.5 kpc (Petrossian, Sahakian and Khachikian 1978). The spectra of the nuclear components are quite identical. Narrow emission lines of [SII] λλ 6717/31, [NII] λλ 6583/48, Hα, [OI] λ 6300, [OIII]λλ 5007, 4959, Hβ, Hγ, [NeIII] λ3869, [OII] λ 3727 are observed in them. The D_1 and D_2, NaI, lines are observed in absorption and are stronger in the weaker component. The bright component displays also a bright short-wave part of the continuous spectrum, while the red parts of their spectra are nearly of the same brightness. The emission lines are also brighter in the bright component; besides the line [OIII]λ 4363 is also noticed in the weak component.

One can presumably confirm the fact that the nuclei of Markarian 266 are quite reminiscent by their physical properties and by their spectra of the objects "twins", referred to above, in paragraph 3, with the basic difference that the objects "twins" form an isolated double system, whereas the components of the nucleus Markarian 266 form a constituent part of the galaxy with diffuse envelope and spiral arms, though poorly developed. This similarity can be of help in solving the problem of the origin of double nuclei.

But most interesting seems to be the discovery of the fact that the components of the nucleus of Markarian 266 rotate, and in opposite directions at that. This is clearly visible by the decline of the spectral lines. Fig. 4b illustrates the portion of the spectrum around Hα, which confirms the above statement. The components of the nucleus rotate at a velocity (in the projection on the celestial sphere) of 40 km/sec (the bright component) and 100 km/sec (the weak component), with their contiguous regions receding from us. The difference of the radial velocities of the components themselves attains about 150-200 km/sec. The component masses calculated by ordinary ways proved to be of the order of 10^9 $M_\odot$ and 6×10^9 $M_\odot$, while the mass-luminosity ratio $f/f_\odot$ is about 1 and 6, for the bright and the weak component, respectively.

It is clear from the above description of the components of the nucleus of Markarian 266 that they are quite similar to super-associations as to their physical characteristics (especially the weak component).

Hence it can be deduced that *one of the forms of activity manifestation is the formation of super-associations, i.e. large groups of young stars.*

DISCUSSION

Thus, relying on an analysis of the morphological and spectroscopic data of galaxies with UV-excess alone, it is evident that most of them manifest themselves as galaxies with one form of activity or another according to Ambartsumian.

Once again, emphasis should be laid on the fact that though the above-considered galaxies differ both in their morphological characteristics and the form of activity, the presence of UV-excess is a trait common to all of them. On the whole their spectra are also similar, though they differ in the particulars and the width of emission lines.

Comparing those data one can arrive at the conclusion that *the physical causes of activity in those galaxies are of a common origin, despite some differences.*

Observational data confirm that the central regions play the basic role in the activity of the galaxies. This activity depends to a much lesser extent on the morphology of the outer parts of the galaxies. On the other hand, the external form of the nucleus is not a factor that can determine the manifestation of one form of activity or another. To illustrate the above statement let us quote the following example. Fig. 5 shows the pictures of Markarian 9, 10, 305 and Kazarian 73. The figure indicates that Markarian 9 and 10 differ sharply in morphology. Markarian 9 is a compact star-like galaxy, while Markarian 10 is a giant spiral galaxy with dimensions about 55 kpc (Khachikian 1970). In spite of this difference, however, they are both Seyfert galaxies of the first type, Syl. If we compare, however, two spiral galaxies with very bright and star-like nuclei, Markarian 10 and Kazarian 73, they differ sharply in activity: Kazarian 73 has narrow but very bright emission lines (see Fig. 5e) while Markarian 10, as noted above, is a Syl type galaxy. (The article about Kazarian 73 is in preparation. The photo and the spectrum of this galaxy are obtained by Kazarian and Petrosian (Byurakan observatory).)

As to the very compact and star-like object Markarian 305, it is not only a non-Seyfert galaxy but lacks emission lines, in general. (Khachikian 1976).

Thus we clearly see from the quoted illustrations that star-like objects, either the nuclei of spiral galaxies or the "bare" nuclei, display quite different forms of activity.

The same is true of the radioemission activity of galaxies with compact nuclei. Let us remind that included in the list of Seyfert galaxies, compiled as early as in 1974 by Weedman and myself, are five Sy1 galaxies which are at the same time radiogalaxies: 3C 120, 3C 227, 3C 2871, 3C 390.3 and PCS2349-01 (Khachikian and Weedman 1974). Subsequently Weedman (1971) supplemented the list of Seyfert galaxies adding three more Sy1 galaxies which are radiogalaxies (4C 296, 4C 35.37, 3C 382).

Detailed spectroscopic investigations of radiogalaxies with emission lines in the spectrum, made by Osterbrock *et al.* (Osterbrock 1976; Kostrero and Osterbrock 1977; Osterbrock, Koski and Phillips 1975, 1976) revealed that those radiogalaxies can be divided into two types as to the form of the spectrum: those with comparatively narrow emission lines, not differing from the optical spectra of Sy2 and those with broad lines, like the spectra of Sy1.

Osterbrock launched a very interesting discussion on the problem at the Copenhagen symposium (Osterbrock 1978). However, we think that it is not diversity but similarity in the spectra of radiogalaxies and Seyfert galaxies that should be sought for. In our opinion, the similarity between them is so striking that one can conclude that radiogalaxies with broad emission lines are Seyfert galaxies of the first type, which, apart from Seyfert activity, possess also radioemission-activity. Similarly: radiogalaxies with relatively narrow lines are Seyfert galaxies of the second type. The reason that many radiogalaxies with an optical spectrum of the Seyfert type are not ascribed to the Seyfert galaxies, lies in the fact that they were originally discovered as radiogalaxies. Those two notions should not be confounded. *Radioemissionactivity and Seyfert activity are two different, possibly independent forms of nuclear activity.* One and the same galaxy can display various forms of activity, as pointed out by Ambartsumian. Perhaps attention should be paid to the following analogy between objects of various classes:

q u a s a r s - o p t i c a l q u a s a r s,

Sy1 galaxies with radioemissionactivity - Sy1 galaxies without that,
Sy2 galaxies with radioemissionactivity - Sy2 galaxies without that.

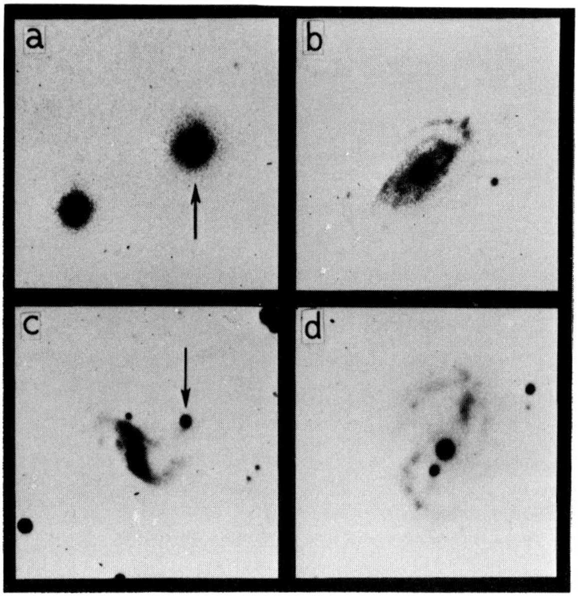

Fig. 5 a. Mark. 3 (5 m Hale telescope)
b. Mark.10 (- " -)
c. Mark.305 + 306 (6 m telescope) arrow shows Mark. 305.
d. Kazarian 73 (6 m telescope)
(ORWO Zu-2, exp. 10 min.)
Photos: Scale: 1 mm ∿ 3"

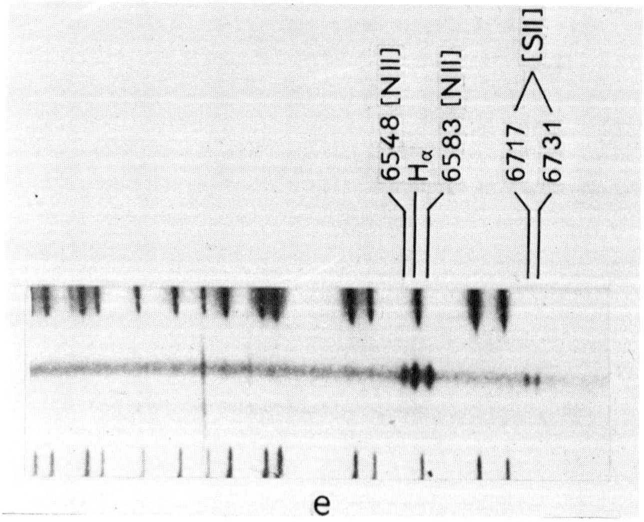

Fig. 5 e. Reproduction of a spectrum of Kazarian 73.
(6 m telescope prime focus spectrograph + I T,
Kodak 103 a-0 exp. 10 min. original disp. ∿ 65 Å/mm

Proceeding from the fact that one form of galactic activity or another does not depend on internal or external morphological peculiarities, one can deduce that some unfamiliar agent, contained in the nuclei of galaxies and for some reasons manifesting itself in different ways, plays a decisive role in the display of activity. This is presumably the super-dense pre-stellar matter which Ambartsumian has constantly been referring to.

At first, after discovering the activity of galactic nuclei many authors expressed the suggestion that this activity is a secondary phenomenon and, in the first place, they took some hypothetical superfast collapse phenomenon.

In the wellknown works by Oort and others was shown that our Galaxy also has had, not long ago (10^7 - 10^8 years ago), activity in the nucleus. From the point of view of the time-scale of galactic evolution the question is literally about the contemporary stage of the evolution of the Galaxy. Meanwhile no phenomena like fast collapse are observed at the present stage of evolution of the Galaxy. Therefore, there is no hope to explain the nuclear activity as a secondary phenomenon.

If it it so, it is necessary to assume extraordinary properties of the nucleus.

If the properties of a magnetoid, a black hole, neutron matter or a super-massive hadron play here a main role - remains an enigma. Only the accumulation of fresh observational data from large telescopes can throw light on this extremely important problem.

REFERENCES

Ambartsumian V.A.: 1958, *La structure et l'evolution de l'univers*, p. 241, Solvay Conference, Ed. R. Stoops, Brussel .

Ambartsumian V.A.: 1962, Trans. IAU XIB, p. 145, Academic Press, London-New York.

Ambartsumian V.A.: 1964, IAU- URSI Symp. No. 20, p. 122, Eds. F.J. Kerr, A.W. Rodgers, Canberra.

Ambartsumian V.A.: 1965, *The Structure and Evolution of Galaxies*, p. 1, Interscience Publ., London.

Ambartsumian V.A., Iskudarian S.G., Shahbazian R.K., Sahakian K.A.: 1963, Soobshch, Byurakan Obs. 33,3.

Arp H.C., Heidmann J., Khachikian E.Ye.: 1974, Astrofizika 10, 7.

Arp H.C., Khachikian E.Ye., Lynds C.R., Weedman D.W.: 1968, Astrophys. J. 152, L103.

Baade W., Minkowski R.: 1954, Astrophys. J., 119, 206.

Bolton J.G., Stanley G.J., Slee O.B.: 1949, Nature 164, 101.

Costero R., Osterbrock D.E.: 1977, Astrophys. J. 211, 675.

Gyulbudaghian A.L.: 1971, Theses of diploma, Yerevan University.

Haro G.: 1956, Bol. Obs. Tonantzintla y Tacubaya 14, 8.

Kalloghlian A.T.: 1971, Astrofizika 7, 521.
Kazarian M.A., Khachikian E. Ye.: 1974, Astrofizika 10, 477.
Kazarian M.A., Khachikian E. Ye.: 1977, Astrofizika 13, 415.
Khachikian E. Ye.: 1968, Astron. J. 73, 891.
Khachikian E. Ye.: 1970, IAU Symp. No 44 p. 160, Ed. D.E. Evans, Reidel Publ. Co., Dordrecht, Holland.
Khachikian E. Ye.: 1972, Astrofizika 8, 529.
Khachikian E. Ye.: 1973, Astrofizika 9, 139.
Khachikian E. Ye.: 1976, Astron. Nachr. 297, 287.
Khachikian E. Ye., Weedman D.W.: 1971a, Astrofizika 7, 389.
Khachikian E. Ye., Weedman D.W.: 1971b, Astrophys. J. 164, L109.
Khachikian E. Ye., Weedman D.W.: 1974, Astrophys. J. 192, 581.
Markarian B.E.: 1967, Astrofizika 3, 55.
Osterbrock D.E.: 1976, Publ. Astron. Soc. Pacific 88, 589.
Osterbrock D.E.: 1978, Phys. Scr. 17, 137.
Osterbrock D.E., Koski A.T., Phillips M.M.: 1975, Astrophys. J. 197, L41.
Osterbrock D.E., Koski A.T., Phillips M.M.: 1976, Astrophys. J. 206, 898.
Petrosian A.R., Sahakian K.A., Khachikian E. Ye.: 1978, Astrofizika 14,69.
Sargent W.L.W.: 1972, Astrophys. J. 173, 7.
Seyfert C.: 1943, Astrophys. J. 97, 28.
Stein W.A., Weedman D.W.: 1976, Astrophys. J. 205, 44.
Weedman D.W.: 1977, Annu. Rev. Astron. Astrophys. 15, 69.
Weedman D.W., Khachikian E.Ye.: 1968, Astrofizika 4, 587.
Weedman D.W., Khachikian E.Ye.: 1969, Astrofizika 5, 113.
Weedman D.W., Khachikian E.Ye.: 1970, Astron. Zirkul. No 591.

MATTER DISTRIBUTION - X-RAY EMISSION FROM CLUSTERS OF GALAXIES

R. Mitchell
Royal Greenwich Observatory, Herstmonceux Castle, Hailsham, East Sussex; Mullard Space Science Laboratory, University College London, Holmbury St. Mary, Dorking, Surrey; and Institute of Astronomy, Cambridge University, Madingley Road, Cambridge, England.

Over six years have elapsed since clusters of galaxies were first established as a class of X-ray sources. This review shall be concerned with describing the developments that have been made in the intervening years, with particular emphasis being given to a discussion of current ideas and problems. In view of the great capabilities of the HEAO-B satellite, soon to be launched, this would appear to be a most appropriate time to take stock of the situation.

The early development of our understanding of X-ray emission from clusters of galaxies has been described in an excellent review by Culhane (1978) and here I shall summarise the main results. Confirmation of the phenomenon of X-ray emission from clusters of galaxies came from the Uhuru satellite after seven rich clusters were found to lie in or near 2U error boxes (Gursky *et al.* 1972). At the time of publication of the 3U catalogue the number of identifications had increased to about twenty (Giacconi *et al.* 1974) with luminosities ranging over 2 orders of magnitude up to about 2.10^{45} ergs s^{-1}. Although of limited spatial resolution the scanning detectors on Uhuru found, for six of these clusters, that the emission was inconsistent with a point source origin. Models of isothermally distributed gas of core radii typically 15', but 25' for Virgo, fitted the data well however (Lea *et al.* 1973; Kellogg and Murray 1974).

An isothermal gas distribution assumes thermal bremsstrahlung as the emission mechanism responsible for producing the X-rays but other evidence supported inverse Compton scattering. There is a high incidence of radio sources within clusters of galaxies, and scattering of the 3K universal radiation field by the relativistic electrons produces X-rays which are characterised by a power law spectrum. Since isothermal X-ray emission gives rise to an exponential continuum, a study of the X-ray spectral properties of clusters of galaxies will distinguish between the two mechanisms. In practice,

Bengt E. Westerlund (ed.), Stars and Star Systems, 123–133.

however, this distinction has proven difficult to observe, mainly because of weak fluxes and because, over the energy range for which many detectors are sensitive ($\lesssim$ 10 keV), the two spectral forms are somewhat similar. Furthermore, bremsstrahlung emission need not arise from an isothermal gas, and Gull and Northover (1975) demonstrated that for adiabatic atmospheres the resulting spectrum resembles a power law. The difficulties are well illustrated in a paper by Kellogg *et al.* (1975) in which spectral data from Uhuru was combined with other observations in an attempt to determine the spectra of the three brightest clusters, Perseus, Virgo and Coma.

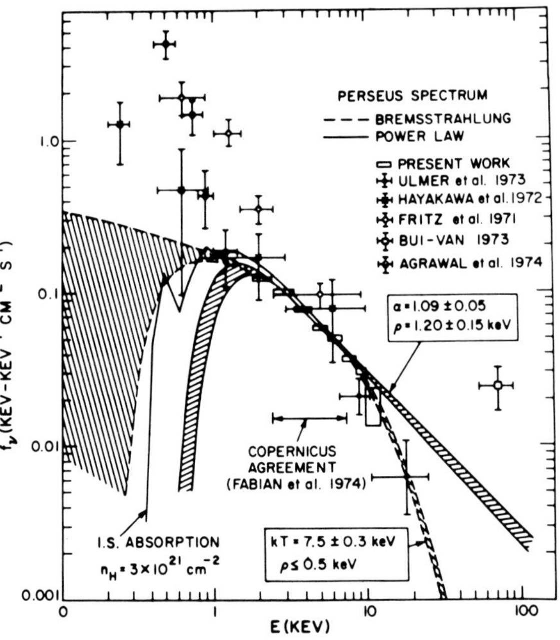

Figure 1. Power law and exponential fits to the Perseus cluster spectrum from data obtained up to 1974.

Spectral observations are not the only means of determining the emission mechanism responsible for X-ray production in clusters of galaxies however. Shortly after the initial discovery, Solinger and Tucker (1972) made a plot of X-ray luminosity, L_x, against velocity dispersion σ_v, for eleven clusters of galaxies and interpreted the relationship observed as being indicative of thermal bremsstrahlung emission. They argued that for inverse Compton scattering in which energy was being released from active galaxies no correlation would be expected between general cluster properties and L_x. The claimed strong dependence of σ_v upon L_x was however inconclusive from the data then available, and the $L_X - \sigma_V$ plot has since been redrawn on numerous occasions with improved parameters or increased statistics. Whether or not a relationship exists other than a rather loose trend still remains uncertain.

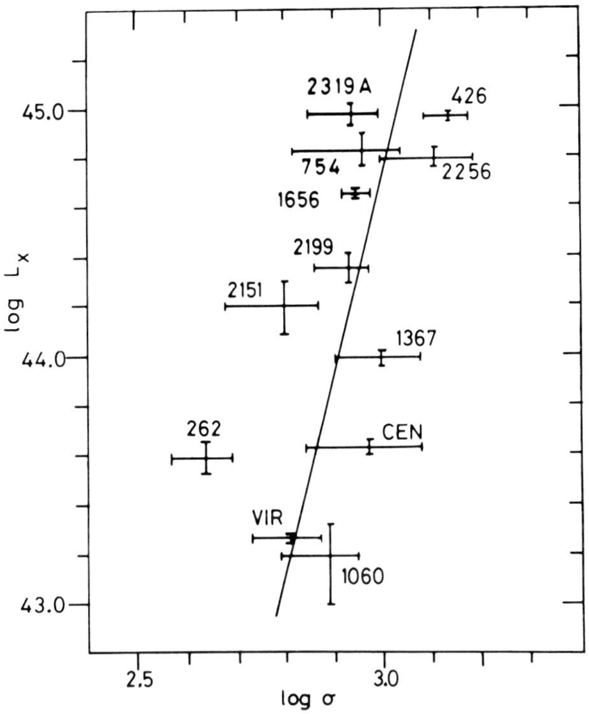

Figure 2. Recent plot of L_X versus σ_V.

Possibly the correlation is poor because it assumes too simple a model. In Virgo, for example, the emission detected by Uhuru appeared to come mainly from M87 rather than the cluster centre. The Copernicus satellite showed that both the Seyfert galaxy NGC 1275 in Perseus and the radio galaxy NGC 4696 in Centaurus are themselves regions of enhanced X-ray emission, each contributing possibly 20 % of the total flux from their respective clusters (Fabian *et al.* 1974; Mitchell *et al.* 1975). The localised region of X-ray emission around NGC 1275 in fact resembles the radio appearance (Wolff *et al.* 1976), and on the larger scale, rather than showing a spherically symmetric distribution, the emission from the Perseus cluster is aligned with the chain of galaxies. The Copernicus satellite was also able to provide information on the low energy ($\lesssim$ 2keV) X-ray emission from clusters. In contrast to the heavily absorbed spectra found in the active galaxies Centaurus A and NGC 4151 which are located outside rich clusters, neither NGC 1275 nor NGC 4696 have measurable column densities, indicating that the X-rays are generated outside the nucleus.

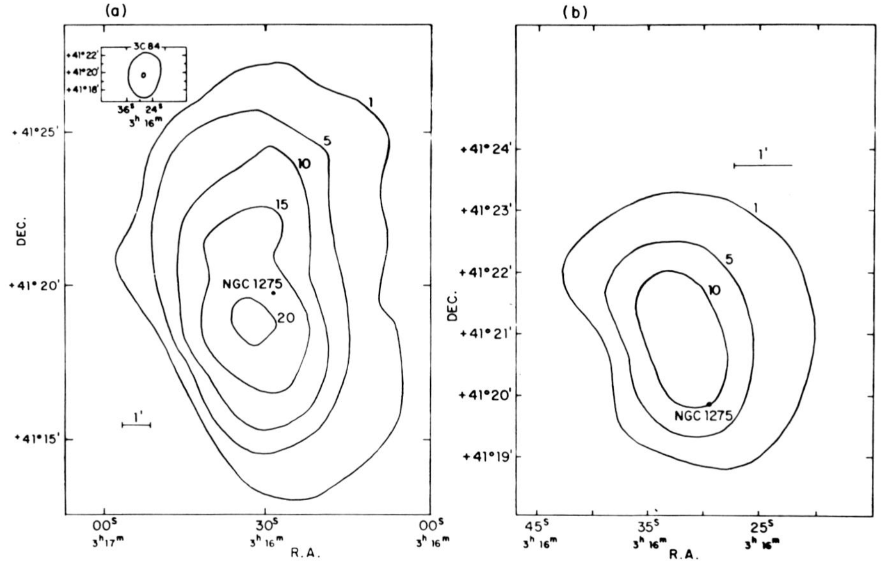

Figure 3. Contours showing X-ray intensity distribution for region around NGC 1275 in the Perseus cluster.

The Ariel V satellite has, more recently, provided a large quantity of both spatial and spectral data. One experiment, the Leicester Sky Survey Instrument (SSI), has succeeded in mapping the sky to generally higher sensitivity than Uhuru. The results have been recently published in the 2A catalogue (Cooke *et al.* 1978) which includes many new sources whose positions correspond with rich clusters. The error boxes are generally smaller than those in the 3U catalogue and have, in some cases, strengthened identifications (e.g. A1795 with 2A 1346 + 266); other clusters previously identified with 3U sources were not detected, however, including A262 for which the Uhuru observations registered extended emission. The updated Uhuru catalogue (4U) has had to be corrected because flux densities of weak sources were overestimated (Forman *et al.* 1978a).

Another detector on Ariel V, the MSSL spectrometer, demonstrated that thermal bremsstrahlung is an important, if not the major, X-ray emission mechanism in clusters by the discovery, in Perseus, of a blend of Fe XXV and Fe XXVI emission lines (Mitchell *et al.* 1976).

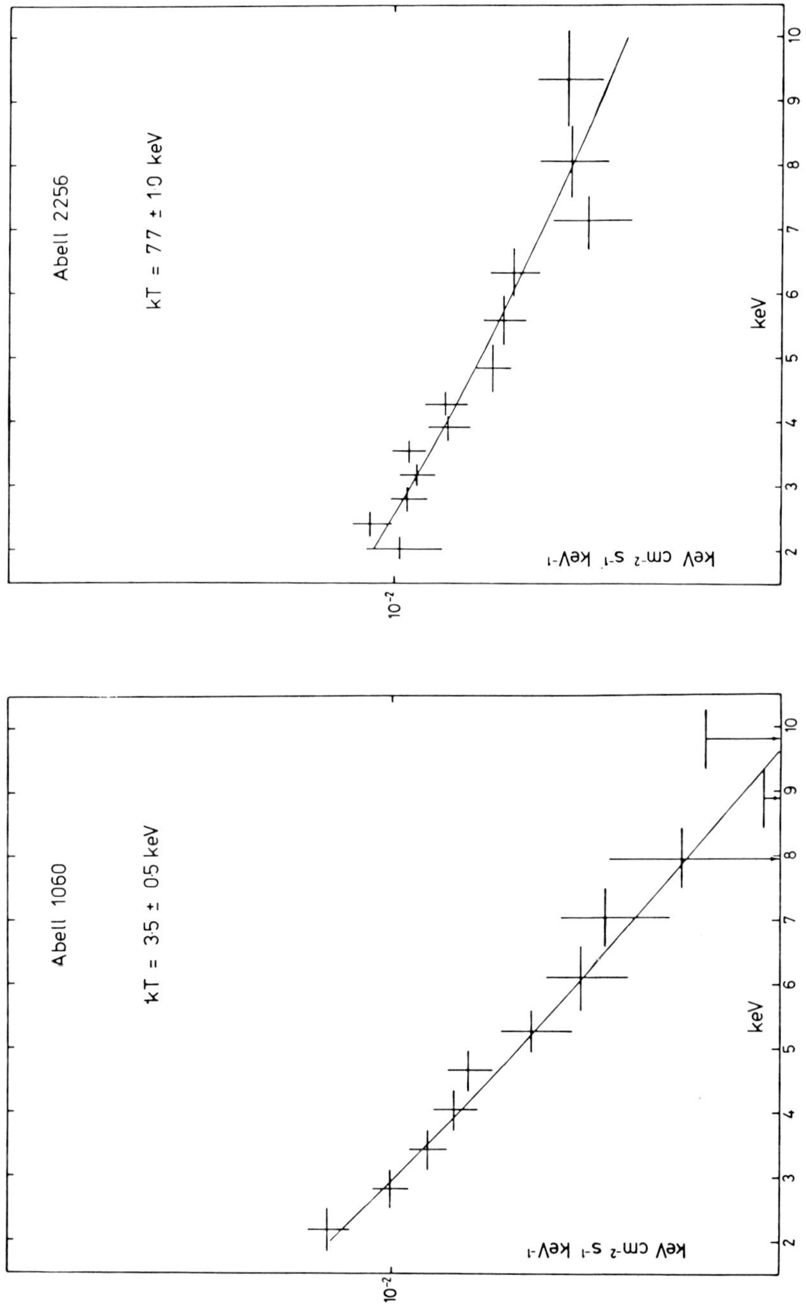

Figure 4. X-ray spectra of two clusters of galaxies from the Ariel V spectrometer.
a) A 1060 b) A 2256

The 6.7 keV feature, which cannot be explained in any non-thermal process, was subsequently observed in the Perseus, Virgo and Coma clusters by the OSO-8 spectrometer, a similar instrument but with a larger field of view (Serlemitsos *et al.* 1977), and in Centaurus by Ariel V (Mitchell and Culhane 1977).

The discovery of an iron line feature meant that the X-ray emission could not arise entirely from a primordial gas. The derived iron abundances however, typically 10-50 % of the cosmic value, did not eliminate the possibility that the line originated in localised regions and that the bulk of the emission had another origin. Evidence for bremsstrahlung being a more general cluster phenomenon also came from the Ariel V spectrometer. Although most of the identified cluster X-ray sources are detectable at only one tenth of the flux value of Perseus, spectra were determined for eight of them and temperatures (kT_x) were derived from the slopes on the assumption of isothermal emission. Correlations relating X-ray and optical properties were tested and, in particular, kT_x was found to be proportional to σ_v^2 (Mitchell *et al.* 1977). Such a result would appear to reconcile if heating was due only to a few active galaxies. OSO-8 extended the number of observed cluster spectra to twenty and, in addition to supporting the correlations derived from the Ariel V results, added a few new ones (Mushotzky *et al.* 1978; Smith *et al.* 1978). These included correlations between emission integral and kT_x and between $\overline{N}_0$, central galaxy density after Bahcall (1977), and kT_x.

A list of cluster X-ray spectra has recently been published based on Uhuru observations (Jones and Forman 1978), and details of about twenty cluster spectra derived from Ariel V observations are in preparation (Mitchell *et al.* 1978). This raises the number of clusters for which X-ray spectral data is available to about thirty although, because of the large fields of view of the Ariel V and OSO-8 spectrometers (3$^\circ\!\!.$5 and 5° FWHM, respectively), the identifications rely mainly upon the 2A and 4U catalogues. We have compared the 2A and 4U error boxes with the positions of the clusters claimed from the identifications, and have assessed each identification according to the sizes and agreement between error boxes, their positions with respect to the cluster, the proximity of other clusters and other criteria such as reported variability. Although many identifications appear sound, a few should be treated with caution. In particular, we find some especially interesting in view of the recent controversy regarding the alleged detection of X-ray emission from superclusters (Murray *et al.* 1978; Pravdo *et al.* 1977, Kellogg 1978).

In several cases the error boxes are either positioned between clusters or include more than one such entity. Both 2A 0255 + 132 and 4U 0254 + 13 are situated between A401 and A399, and neither cluster is located within the 90% confidence regions. A similar situation occurs for 2A 1508 + 062 and the clusters A 2029 and A 2033.

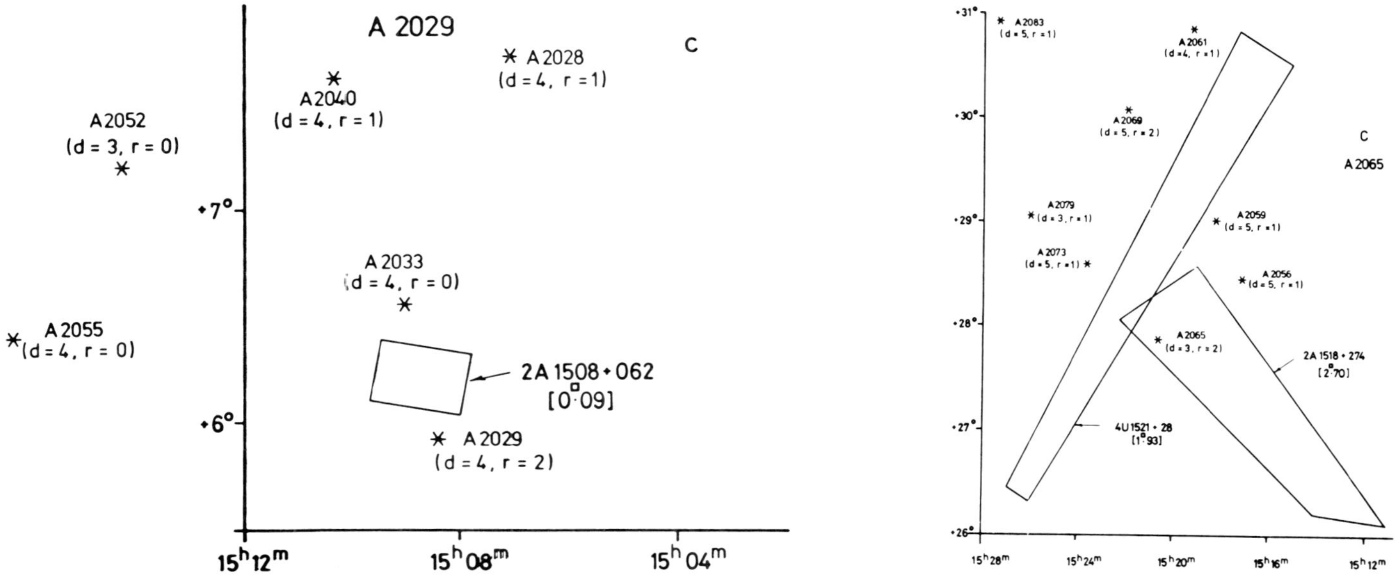

Figure 5. 2A and 4U sources identified with clusters of galaxies.
a) 2A 1508+062 (A 2029 ?) b) 2A 1518+274/4U 1521+28 (A 2065).

The large error boxes for 2A 1518 + 274 and 4U 1521 + 28 are in a region containing many distant class 4 and 5 clusters. Rood (1976) has demonstrated that about 40 percent of clusters out to Abell distance class 4 may occur in superclusters, and so it is possibly not surprising that among the X-ray cluster sources several are associated with these superclusters. Whether X-ray emission originates within superclusters or whether the effect is simply an artifact of inadequate resolution remains to be seen.

Another large scale effect has been noted recently by Forman *et al.* (1978) who demonstrated that the measured flux values are greatest for detectors with the largest fields of view. This result is supported by the flux values obtained from the Ariel V spectrometer which are, in general, higher by a factor of between 1 and 2 than the SSI values. In addition, there is new evidence which implies that the size of the emission region increases with increasing energy. This appears to be true for both the Perseus cluster (Ulmer and Jernigan 1978) and for A478 (Schnopper *et al.* 1977), two clusters which have been studied with the instruments on the SAS-3 satellite, and for Virgo, which was investigated using a two dimensional imaging X-ray telescope flown on board a sounding rocket (Gorenstein *et al.* 1977). The data in all three cases are consistent with a point source component centred on a dominant galaxy in addition to the more diffuse emission. For Virgo, the active galaxy, M87, is away from the cluster centre, and observations with the Ariel V spectrometer show that, whereas the bestfit position for the source of the soft ($\lesssim$ 5 keV) X-rays is coincident with M87, the harder energy flux ($\gtrsim$ 8 keV) originates from a region much nearer the cluster centre (Davison 1978).

The recent observations with improved spatial and spectral resolution indicate that X-ray emission from clusters of galaxies cannot be explained by the simple models extensively used until now. Indeed, even for Coma, considered to be a rich and regular cluster, the X-ray appearance is of a clumpy and non-spherically symmetric system (Gorenstein *et al.* 1978). Energy separation effects similar to those seen in Virgo have been observed by Ariel V in several other clusters, including A401, A1367 and A2256 (Mitchell *et al.* 1978). In Virgo, a second spectral line feature has been discovered with the low energy detector on HEAO 1. (Lea *et al.* 1978). The feature, at 1.1 keV, is consistent with a blend of iron lines and corresponds to an iron abundance of about one third the cosmic value.

In conclusion, I shall mention a recent model for X-ray production in clusters of galaxies proposed by Lea and Holman (1978). The model is attractive because it attempts to explain the X-ray emission as thermal bremsstrahlung but with radio sources playing an important role. Essentially, material processed in stars is ejected from galaxies and, together with any primordial gas in the neighbourhood, is heated by collisions with ejecta from other galaxies. The cluster central density increases and cooling occurs at a rate governed largely

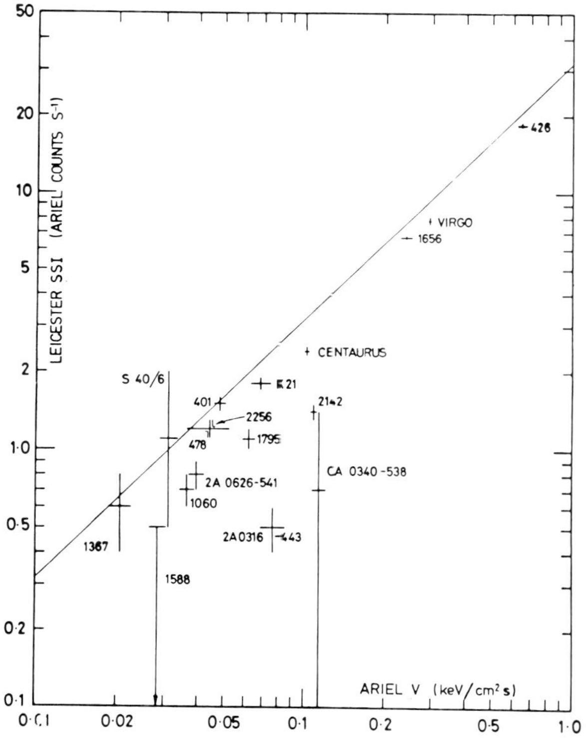

Figure 6. Comparison of fluxes measured by instruments with different fields of view. Vertical axis gives value from SSI; horizontal axis gives values from Ariel V spectrometer. Solid line shows expected relation for constant point sources.

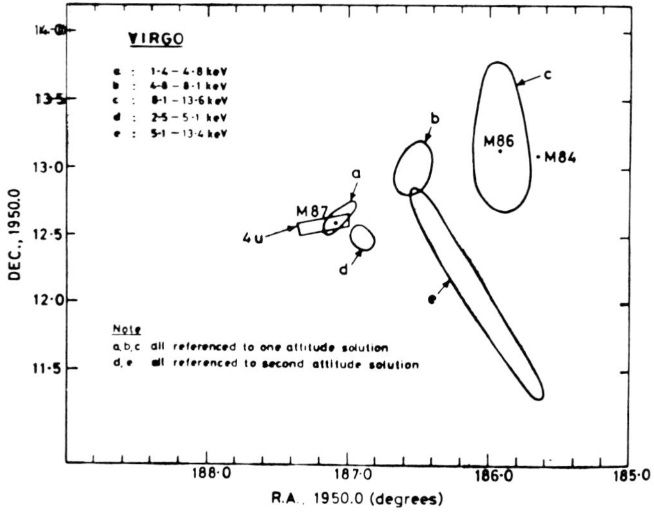

Figure 7. Energy separation effects in Virgo. Region of soft X-ray emission is coincident with M 87.

by the cluster parameters, but probably within a Hubble time. In order to maintain X-ray emission a secondary heating mechanism is necessary, and Lea and Holman propose that this is by the relativistic electrons which are responsible for the radio synchrotron emission. The situation would appear to be delicately balanced, for if the intracluster medium is of low density and if the radio source is particularly powerful, winds may develop which will dissipate tha gas and result in a lower X-ray luminosity. Alternatively, for a high density medium both a high value of L_x and a steep low frequency radio spectrum are predicted. Perhaps the correctness of the model will be decided from the data which should come from HEAO - B.

REFERENCES

Bahcall, N.A.: 1977, Astrophys. J. 217, L77.
Cooke, B.A., Ricketts, M.J., Maccacaro, T., Pye, J.P., Elvis, M., Watson, M.G., Griffiths, R.E., Pounds, K.A., McHardy, I., Maccagni, D., Seward, F.D., Page, C.G. and Turner, M.J.L.: 1978, Mon. Not. R. astr. Soc., 182, 455.
Culhane, J.L.: 1978, Q.J. R. astr. Soc. 19, 1.
Davison, P.J.N.: 1978. Mon. Not. R. astr. Soc. 183, 39^{p}.
Fabian, A.C., Zarnecki, J.C., Culhane, J.L., Hawkins, F.J., Peacock, A., Pounds, K.A. and Parkinson, J.H.: 1974, Astrophys. J. 189, L59.
Forman, W., Jones, C., Cominsky, L., Julien, P., Murray, S., Peters, G., Tananbaum, H. and Giacconi, R.: 1978, submitted to Astrophys. J. Suppl.
Forman, W., Jones, C., Murray, S.S. and Giacconi, R.: 1978, preprint.
Giacconi, R., Murray, S., Gursky, H., Kellogg, E., Schreier, E., Matilsky, T., Koch, D., Tananbaum, H.: 1974, Astrophys. J. Suppl. 27, 37.
Gorenstein, P., Fabricant, D., Topka, K., Tucker, W. and Harnden, F.J.: 1977, Astrophys. J. 216, L95.
Gull, S.F. and Northover, K.J.E.: 1975, Mon. Not. R. astr. Soc. 173, 585.
Gursky, H., Solinger, A., Kellogg, E.M., Murray, S., Tananbaum, H. and Giacconi, R.: 1962, Astrophys. J. 173, L99.
Jones, C. and Forman, W.: 1978, Astrophys. J., in press.
Kellogg, E.M.: 1978, Astrophys. J. 220, L63.
Kellogg, E., Baldwin, J.R. and Koch, D.: 1975, Astrophys. J. 199, 299.
Kellogg, E. and Murray, S.: 1974, Astrophys. J. 193, L57.
Lea, S.M. and Holman, G.D.: 1978, Astrophys. J. 222, 29.
Lea, S., Mason, K., Reichert, G., and Riegler, G.: 1978, preprint.
Lea, S., Silk, J., Kellogg, E. and Murray, S.: 1973, Astrophys. J. 184, L105.
Mitchell, R.J., Charles, P.A., Culhane, J.L., Davison, P.J.N., and Fabian, A.C.: 1975, Astrophys. J. 200, L5.
Mitchell, R.J. and Culhane, J.L.: 1977, Mon. Not. R. Astr. Soc. 178, 75^{p}.
Mitchell, R.J., Culhane, J.L., Davison, P.J.N. and Ives, J.C.: 1976, Mon. Not. R. astr. Soc. 176, 29^{p}.
Mitchell, R.J., Dickens, R.J., Bell-Burnell, S.J., and Culhane, J.L.: 1978, in preparation.

Murray, S.S., Forman, W., Jones, C. and Giacconi, R.: 1978, Astrophys. J. 219, L89.
Mushotzky, R.F., Serlemitsos, P.J., Smith, B.W., Boldt, E.A., and Holt, S.S.: 1978, preprint.
Pravdo, S., Mushotzky, R., Becker, R., Boldt, E., Holt, S., Serlemitsos, P. and Swank, J.: 1977, Nature 270, 158.
Rood, H.J.: 1976, Astrophys. J. 207, 16.
Schnopper, H.W., Delvaille, J.P., Epstein, A., Helmken, H., Harris, D.E., Strom, R.G., Clark, G.W., and Jernigan, J.G.: 1977, Astrophys. J. 217, L15.
Serlemitsos, P.J., Smith, B.W., Boldt, E.A., Holt, S.S. and Swank, J.H.: 1977, Astrophys. J. 211, L63.
Smith, B.W., Mushotzky, R.F., and Serlemitsos, P.J.: 1978, preprint.
Solinger, A. and Tucker, W.: 1972, Astrophys. J. 175, L107.
Ulmer, M.P. and Jernigan, J.G.: 1978, Astrophys. J. 222, L85.
Wolff, R.S., Mitchell, R.J., Charles, P.A. and Culhane, J.L.: 1976, Astrophys. J. 208, 1.

COLOUR SYSTEMS AND MODELS OF STELLAR ATMOSPHERES

B. Gustafsson
Uppsala Universitets Astronomiska Observatorium
Uppsala, Sweden

ABSTRACT

The theoretical fluxes from stellar model atmospheres and synthetic-spectrum computations may be convolved with the sensitivity functions of photometric systems in order to obtain theoretical colours. These colours may be used for various purposes, such as

i) determining fundamental parameters and other physical properties of stars,
ii) suggesting or choosing photometric systems for specific purposes,
iii) analysing the properties and capabilities of different photometric systems.

Examples of applications of these, and other, types are given. The power and the shortcomings of the synthetic-colour technique are also illustrated and discussed.

1. WHAT ARE SYNTHETIC COLOURS?

Several years ago Anne Underhill made a statement, often cited by stellar astronomers in difficulties: "The problem of fitting a series of beautiful internally-consistent models to honest to goodness real stars that are up there, is horrible". This talk will deal with certain aspects of this horrible, though important and even quite interesting, problem: Starting from our internally, more or less consistent model atmospheres, we compute their spectra in great detail, multiply the spectra with the sensitivity functions of various photometric systems and confront these computed or *synthetic* colours with observed colours of real stars. What do we learn from such an effort? I shall try to give some answers to that question.

First, however, by discussing Fig. 1 I would like to remind you about some fundamentals of the procedure outlined. In constructing the model atmospheres we first make certain, often questionable *basic assumptions*, such as plane-parallel stratification, hydrostatic

Bengt E. Westerlund (ed.), Stars and Star Systems, 135–153.

equilibrium and LTE. (The main reason why oversimplifying assumptions are made is not the naive optimism of the astrophysicists but the absence of a complete theory for stellar atmospheres and of relevant physical data and numerical methods.) Guided by theoretical physics we are then able to formulate a number of equations. In order to solve these we need, often quite extensive, physical data, e.g., on various cross sections. In particular, the computation of detailed synthetic spectra requires considerable data - in our own work on late-type giants we have had to include several hundreds of thousands of spectral lines in the synthetic-spectrum computations. And yet these computations were made assuming LTE - a detailed non-LTE treatment of the interaction of the atoms and molecules with the radiation would have required much more extensive data.

We also need data specifying what star the model is intended to resemble. These latter data are often called *fundamental parameters* and are generally specified as effective temperature, surface acceleration of gravity and chemical abundances. There may be additional parameters entering at this stage, like the mixing-length parameter and the microturbulence parameter; quantities beyond any well-founded physical interpretation but necessary to introduce since the theory of the hydrodynamics of stellar atmospheres is not sufficiently developed. However, the uncertainties introduced by these *ad hoc* parameters, and their underlying erroneous assumptions, do not seriously affect most applications of the synthetic-colour technique, like the results to be discussed below.

The equations are reformulated in a suitable way and are fed into a computer with the data, and out comes a model atmosphere (if a proper numerical method was chosen to solve the equations).

When the spectrum of the model is compared with good spectra of the corresponding star, discrepancies almost inevitably show up. It is always tempting, and often correct, to ascribe such discrepancies to errors in the fundamental parameters assumed for the star. Therefore, some adjustment of these parameters may be made such that a better fit is obtained. The adjusted parameters are often regarded as new, and better, determinations of, for instance, the chemical composition of the star. It as also often tempting, and sometimes correct, to ascribe such discrepancies to errors in the basic assumptions or in the physical theories or data used in the computations. It is a thrilling challenge in this branch of science to attempt to judge what the really safe conclusions from comparisons of this type are in any specific case. One necessary way of making such judgements possible is to improve the physical theory of stellar atmospheres to allow for, e.g., mass flows, inhomogeneities, departures from LTE in complex atoms and molecules, etc. Another, complementary, way is to compare the observed spectra with the spectra of the most detailed models we can construct today and to keep in mind that the inconsistencies that turn up in the comparison with observations should not be swept under the carpet (which easily happens) but ought to be analysed systemati-

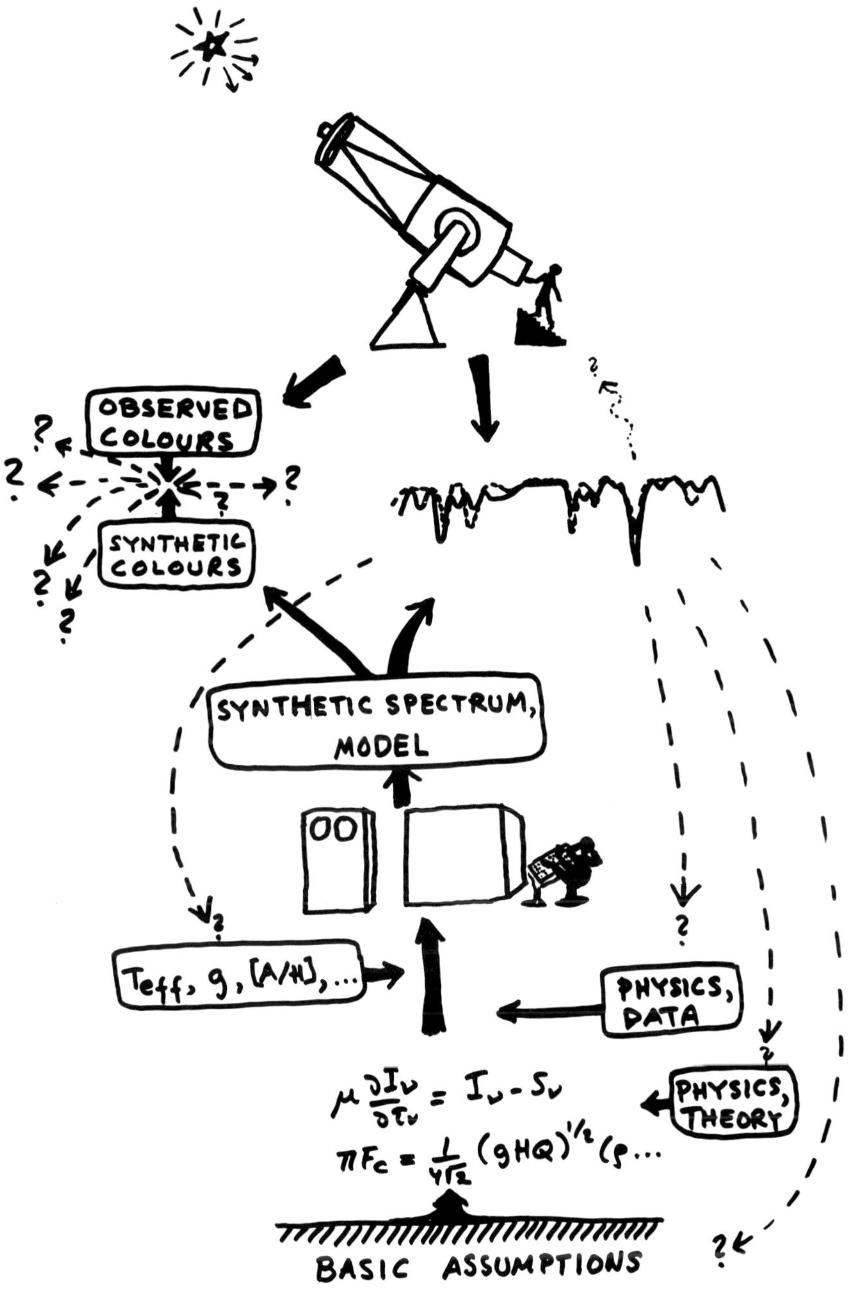

Figure 1. A very schematic picture of how model atmospheres for stars are computed and compared with observations. The present paper mainly deals with the questions indicated in the upper left part of the figure.

cally (which is often rather involved). Also, we should remember that the fundamental parameters, such as abundances, obtained in these comparisons should not be taken quite seriously.

In this talk I shall discuss results from comparisons between models and stars in quite low resolution. The high-resolution theoretical spectra have been multiplied with the sensitivity functions of photometric systems. Into these sensitivity functions we include transmission functions of the filter, the telescope and the terrestrial atmosphere, and the response function of the detector. After an integration over wavelength we obtain a magnitude measure and colours can be derived as differences between magnitudes. Before these colours can be compared with observed ones we have to establish the zeropoint corrections necessary for the theoretical colours so that we can insert them into the observational system. The zeropoints can be derived by adopting a model with certain parameters for a well-observed star. This standard star may be of early spectral type, which is favourable due to the relatively simple spectrum in the visual, or it may be similar to the stars being investigated, in order to minimize the effects or errors in the model atmospheres.

Several investigators have computed synthetic colours and compared them with observations in extensive studies. Some of the most recent ones are tabulated in Table 1.

Table 1. Recent studies on synthetic colours and comparisons with observations

Author(s)	Interval in T_{eff}(K)	Interval in log g	Interval in [A/H]	Colour systems
Bell and Gustafsson (1975), Bell and Gustafsson (1978), Gustafsson and Bell (1978).	4000-6000	0.75-3.0	-3.0-0.0	UBVR, *uvby*, Geneva, *gnkmf*, DDO, c_e-g_e, JML, Oke scans
Lub and Pel (1977)	5500-8000	1.0-4.5	-2.0-0.0	Walraven *VBLUW*
Peytremann (1975)	5000-8500	2.0-4.5	-1.0-0.0	Geneva
Relyea and Kurucz (1978), Buser and Kurucz (1978), Kurucz (1978).	5500-50000	0.0-4.5	-2.0-0.0	UBV, *uvby*

In the present talk I shall mainly concentrate on colours computed by Rogėr Bell and me for a grid of red giant model atmospheres, computed by us and collaborators a couple of years ago (Gustafsson *et al.* 1975, Bell *et al.* 1976a). The models include metal and molecular line blanketing and mixing-length convection. They cover spectral types from G0 to M0 and metal abundances from solar to 1/1000 of the solar (here denoted [A/H] = 0.0 and -3.0, respectively). The colours of many different systems were computed - here we shall confine the discussion to the UBVRI, *uvby*, Geneva 7 colour system, the DDO and the Brorfelde *gnkmf* systems and finally the Uppsala $c_e - g_e$ system. The computed colours are given and extensively discussed in papers in press (Bell and Gustafsson 1978, Gustafsson and Bell 1978, called *Paper I* subsequently).

II. HOW GOOD ARE SYNTHETIC COLOURS?

A necessary but probably not sufficient condition for a model atmosphere to be realistic is that the model reproduces all accurate observations of the emitted spectrum from the star. In a first attempt to answer the question posed by the heading one may compare the synthetic spectra with observed high-dispersion spectrograms and accurate low-resolution scans and similar data. We have performed numerous comparisons of this type for red giants and the agreement must be considered good or satisfactory. Several examples are given in Paper I.

A second test is to plot synthetic two-colour diagrams in the various photometric systems and investigate whether the model atmospheres line up along the observed sequences in those diagrams. We have used this method for the giants but we have also tried a more conclusive approach: we selected a sample of giant stars with comparatively well determined fundamental parameters and well distributed in the T_{eff}-log g - [A/H] space (to find such stars, and trust their parameter estimates, are of course major difficulties of this approach). We then used our grid of synthetic colours to predict the colours of the stars and formed the differences Δ = observed - predicted colour. These differences were often gratifyingly small but for *two* classes of colours they were not. For the ultraviolet colours, like U-B or the Strömgren c_1 index, the metal-rich models were too bright in the ultraviolet. This discrepancy is shown in Fig. 2 and is obviously a function of the metal abundance. In fact, some of this discrepancy is present even for the Sun, as is shown in Paper I.

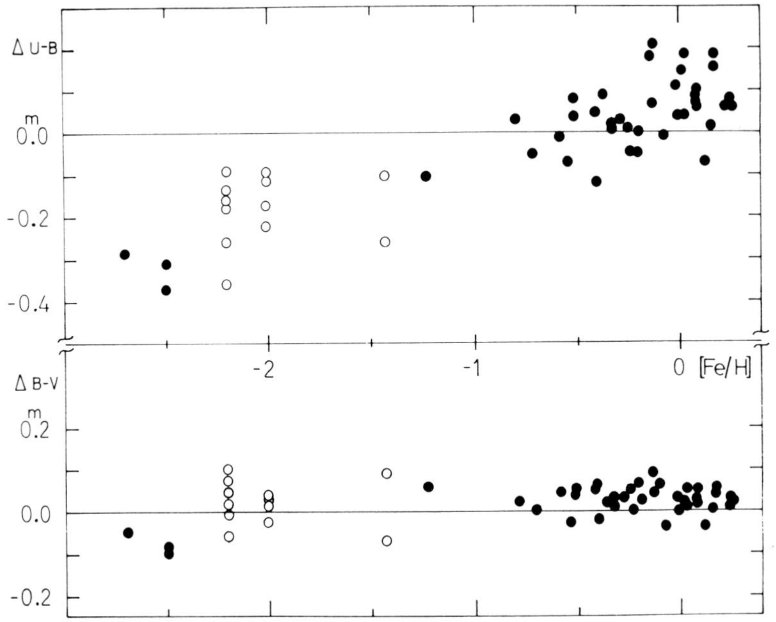

Figure 2. The differences, Δ, between observed and calculated U-B and B-V colours are plotted versus [Fe/H]= [A/H]. The points represent field giants and the open circles denote giants in globular clusters.

The most probable explanation for this ultraviolet discrepancy is an incompleteness in our list of spectral lines, although we include hundreds of thousands of lines in these computations. Probably a veil of many very faint lines is the reason for the discrepancy. (However, there are also other possible explanations, which are discussed in Paper I.) This phenomenon, often referred to as "the unknown opacity source of the Sun" obviously is worse for the Population I red giants. Any hypothesis of its nature should explain this fact and the increase in Δ with [A/H].

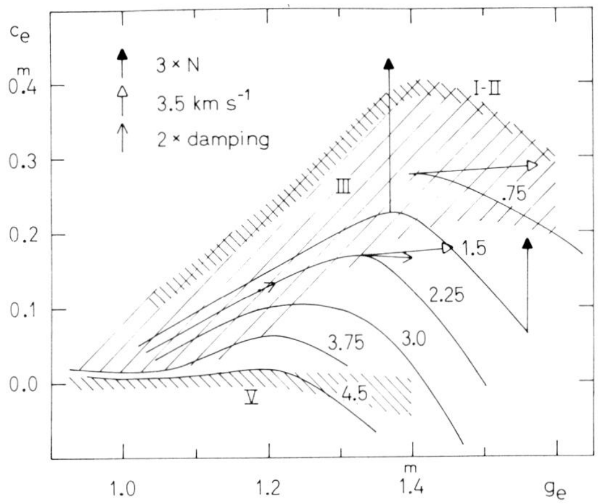

Figure 3. The loci of solar-abundance models with different surface acceleration of gravity in the c_e- g_e diagram of the Uppsala system (continuous curves). The relevant logarithmic gravity is indicated for each curve. The regions occupied by Population I stars of different luminosity classes are also indicated (dashed). The effects of changes in the nitrogen abundance, in the Doppler Broadening Velocity (corresponds to the microturbulence parameter) and in the damping parameter (reflecting uncertainties in the theory of line broadening) are shown by arrows. Typical logarithmic gravities for K-type dwarfs are 4.5, for K giants 3.0 - 1.5 and for bright giants and supergiants 1.5 - 0.5. Obviously, the giants and supergiants are above their computed locations in the diagram (from Paper I).

The second discrepancy we have found occurs for the numerous indices measuring the strength of the famous (0,1) band of the violet CN system with band head at 4215 Å and other CN bands belonging to the same sequence Fig. 3 illustrates the situation in the two-colour diagram of the Uppsala c_e - g_e system. When we compare the predicted locations with the observed ones we find that the real luminosity sensitivity of the c_e index is much greater than predicted. Similarily, the observed metal-abundance sensitivity of the c_e index is considerably greater than predicted by the models. We think that the most important explanation for this discrepancy is the fact that most giants have nitrogen-enriched atmospheres, as a result of mixing of CNO processed material to the surface. This mechanism was discussed by Iben already in the sixties and later by Hejlesen (1976). Recently Lambert and Ries (1977) have, at least partially, verified the existence of this enrichment by detailed spectroscopic analyses.

Both these discrepancies illustrate the fact that we can improve our knowledge of the physics of stellar atmospheres and stars by computing synthetic colours and comparing them with observations.

One could ask whether other, even more fundamental, points in the physics of stellar atmospheres, such as inhomogeneities, hydrodynamic phenomena, departures from LTE and the existence of chromospheres and circumstellar shells may be studied by confronting synthetic colours with observations. The answer is that *some* of this may be possible, *provided that additional, theoretical or observational information of high quality is present.* Do not believe in sensational conclusions based on colours alone!

III. WHAT CAN SYNTHETIC COLOURS BE USED FOR?

A first application of synthetic colours - to study the basic physics of stellar atmospheres and the extent to which our models are realistic - was illustrated and discussed above. A second, more commonplace application is to use the colours to *derive fundamental parameters and establish calibrations.* The synthetic colours may be valuable for many purposes of this type but especially in situations when empirical calibrations are poor or lacking, for instance due to the fact that nearby bright calibration stars are poor or lacking. Three examples of such applications are given below (due to my love of ease chosen from work in which I have been involved):

1) Many red giant stars in the very metal-poor globular clusters M 92 and NGC 6397 seem to show comparatively weak G bands. In fact, Roger Bell, Bob Dickens and I have recently shown that the luminous giants high up along the giant branch of the clusters have G bands indicating a carbon deficiency of a factor of ten below the metals, which in turn are down by a factor of a hundred relative to the Sun. It is not quite certain, though probable, that this is a true abundance effect - it might possibly also be a consequence of departures from LTE. Anyhow, it is of interest to know whether such (apparent) deficiencies also show up for the fainter giants, further down along the giant branch. These stars are so faint that we have no spectra of them. But we can use the *C(42-45)* colour of the DDO narrowband photometry as a G band index, calibrated theoretically for varying carbon abundances. From Fig. 4 we find that the faint giants do not show any strong G-band weakening, thus indicating that the effect for the luminous giants is intrinsic to the stars, and not a general abundance effect for the cluster.

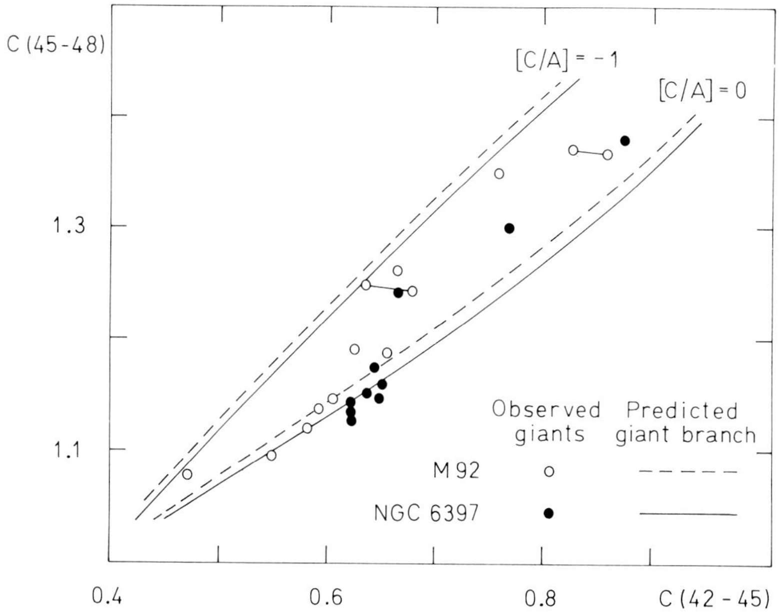

Figure 4. The locations of stars of the globular clusters M 92 and NGC 6397 in the *de-reddened C(45-48) - C(42-45)* diagram of the DDO photometry. Also shown are the predicted giant branches for the two clusters and the corresponding predictions if the carbon abundance is assumed to be reduced by a factor of ten relative to that of iron (adapted from Bell, Dickens and Gustafsson 1978).

2) The Draco dwarf spheroidal system in the outskirts of the Galactic halo was suggested some years ago by Hartwick and McClure (1974) to have a metal abundance of about 1/1000 of the solar, i.e., the system was claimed to be more metal poor than any studied globular cluster. This statement was based on DDO photometry and was interesting, if true, since it gave important restrictions on models for the formation and early evolution of our galaxy.

Roger Bell and I rediscussed the DDO observations, on the basis of our theoretical calibration, and found that the Draco system has the more normal metal abundance of about 1/100 of the solar (Bell and Gustafsson 1976). This higher metal abundance was recently confirmed in a study by Zinn (1978).

3) The relatively young cluster NGC 2209 in the outer regions of the Large Magellanic Cloud was recently suggested by Gascoigne *et al.*(1976) to be significantly more metal poor than stars in the central part of the Cloud. The suggestion was based on UBV photometry for many and DDO photometry for a few cluster stars. This is an example of a

situation where straight-forward use of calibrations based on stars in our Galaxy is dangerous, since we simply do not know of any young massive metal-poor stars here. We found that if the mass differences were properly considered between metal-poor stars in the Galaxy and in the Cloud, the result of the metal-abundance estimation changes to a value between 1/3 and 1/10 of the solar one (Gustafsson, Bell and Hejlesen 1977). This is not significantly different from what is obtained for stars in the central part of the Cloud.

One may ask whether synthetic colours will also play the main role when calibrating photometric systems for stars well represented in the solar neighbourhood. I think careful *empirical* calibrations will still be of the uttermost importance, especially in calibrating abundance indices. One example of this is the calibration by Nissen and Gustafsson (1978) of the Strömgren δm_1 index in terms of [Me/H] for F dwarfs, where the strength of a group of very weak metal lines was measured in many dwarf spectra. Accurate [Me/H] values were derived from a model-atmosphere analysis of these measures, and the [Me/H] values were then found to correlate very well with the δm_1 indices. The resulting calibration is at present superior to what can be expected from a direct theoretical attack. One reason for this is that the m_1 indices, but not the weak lines Nissen measured, are sensitive to the microturbulence parameter, the value of which theories of dynamics in stellar atmospheres cannot predict accurately. However, synthetic colours are still of great value in the calibrations, for instance when establishing the temperature scales and when choosing *forms* of the calibration functions -- it is thus clear from Fig. 3 that an assumed straightforward linear calibration for relating the $c_e - g_e$ indices to the fundamental parameters would be quite dangerous.

A very interesting application of synthetic colours in future will be using them in *analyses of integrated light from galaxies* for synthesizing their stellar populations and chemical compositions. An early example of an application of this type is the study of the Mg index at 5175 Å in old galaxy populations by Mould (1978). Obviously, the synthetic colour approach will sometimes be the only one possible, since single stars with the properties of the stars in the galaxy may be out of reach for observations. However, such applications will often require very reliable model atmospheres and synthetic colours. This thus motivates considerable efforts today, even for detailed investigations of the discrepancies between theory and observations for the comparatively familiar stars in the solar neighbourhood.

A different application of synthetic colours is to use them to *suggest or choose photometric systems for specific purposes*. One example of this is a study Ardeberg and I made of the possibilities for surveying horizontal-branch stars of Intermediate Population II in different regions of the Galaxy (Gustafsson and Ardeberg 1978). It would be valuable to be able to find such stars since they are of interest in themselves (being in late evolutionary stages) and also

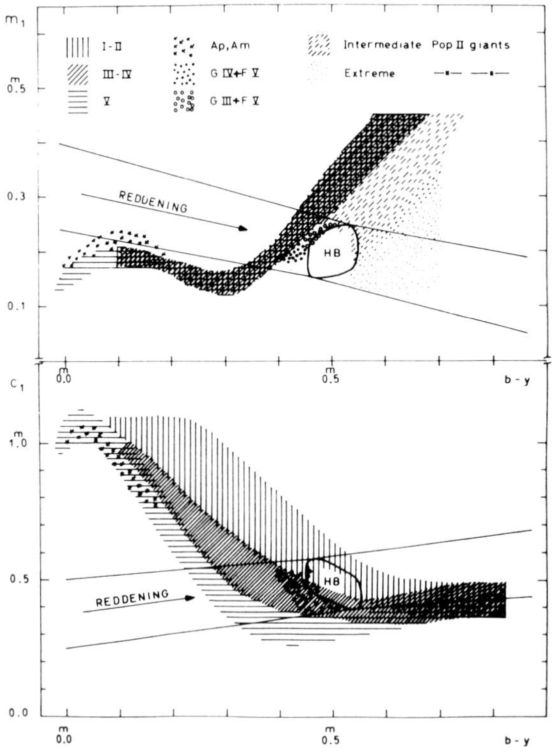

Figure 5. The domains of horizontal-branch stars of Intermediate Population II (marked "HB") and the stars of other types, mainly of Population I, in the $m_1-(b-y)$ and $c_1-(b-y)$ diagrams of the Strömgren *uvby* photometry. The strip along which are located such stars as might be shifted into the HB domain by interstellar reddening is indicated. This figure and the discussion by Gustafsson and Ardeberg (1978, from which the figure is taken) shows that the possibilities are good for using *uvby* photometry to find red field HB stars and separate them from stars of other types when the interstellar reddening in E(B-V) is known whitin about $0^m.10$.

in galactic-structure work. Our synthetic colours showed that the standard *uvby* system was quite appropriate for this purpose. Observations of stars in the metal-rich globular cluster 47 Tuc later verified this conclusion (cf. Fig. 5).

Synthetic colours may also be used in *analyses of photometric systems to find out what is actually being measured.* This is a very important and natural application of the technique. Here two examples of such an application will be given; more examples may be found in the papers listed in Table 1 and elsewhere in the literature.

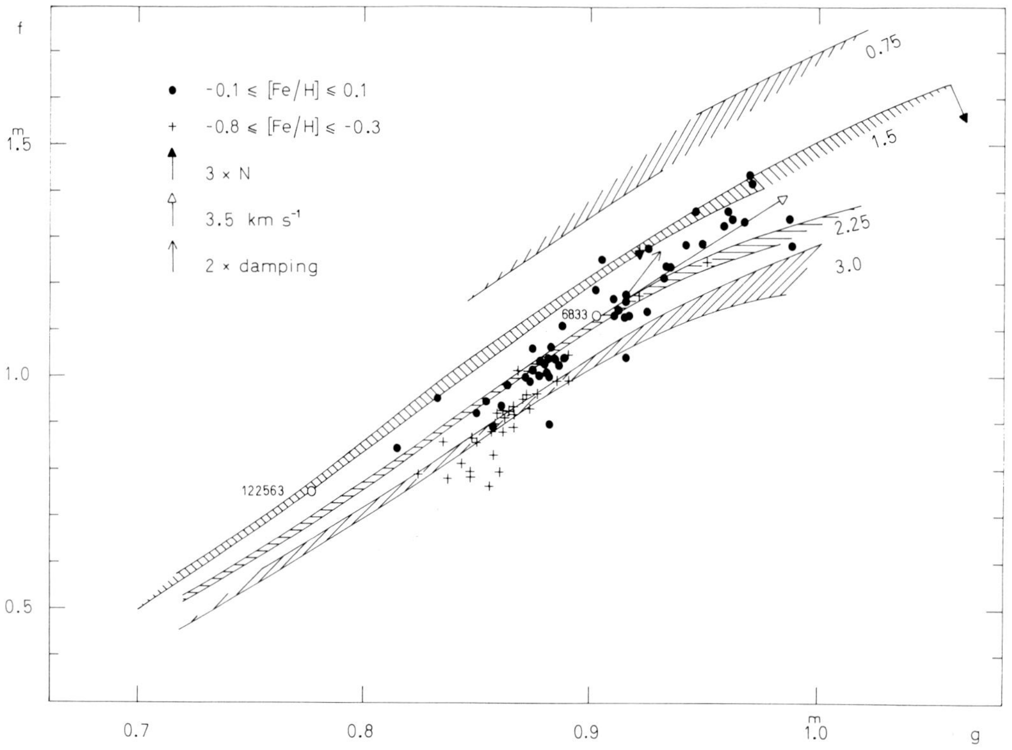

Figure 6. The loci of model atmospheres for red giants in the *f*-*g* diagram of the narrow-band *gnkmf* photometry. *g* measures the break around the G band, i.e. it is first of all a temperature measure; *f* is defined in the text. For each gravity (indicated to the right in the figure) all models with $-1.0 \leq$ [A/H] ≤ 0.0 are located within the corresponding dashed region in the diagram. The separation of models with different gravities in this diagram is clearly seen.

The effects of changes in nitrogen abundance, Doppler Broadening Velocity and damping parameter are indicated as arrows. Points representing observations of stars of two different metal-abundance groups are also plotted. (Figure from Paper I.)

In the Danish *gnkmf* system the *f* index, defined as $m_{4057} - m_{4973}$ with bandwidths of about 80 Å, is an excellent gravity measure for

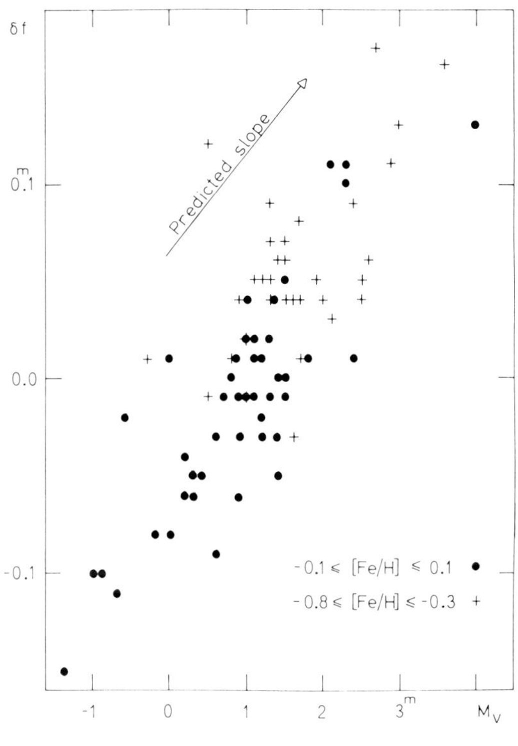

Figure 7. Values of δf (defined in the text) versus $M_V(K)$ for red giants belonging to two different abundance groups. The theoretical slope of the relation between δf and M_V is indicated. (Figure from Paper I.)

G8 - K2 giants, at least theoretically, as is shown in Fig. 6. In this figure we defined a mean line with a mean slope, representing models with equal surface gravity. We measured the departure δf in f from that mean line for a number of stars with absolute-magnitude estimates from the width of the K-line emission ($M_V(K)$). We found that this δf correlated very well with $M_V(K)$ and that the slope of this correlation was close to that predicted (Fig. 7). Thus, the power of this index for determining gravities of red giants is empirically verified. What are the reasons for this gravity sensitivity? We have found that there are two main reasons, of approximately equal importance. First, the smaller the gravity, the stronger the line blocking in the 4057 Å band becomes due to the lower electron pressure and thus smaller continuous H^- opacity. Secondly, the importance of the Rayleigh scattering, greater towards short wavelengths, increases relative to H^- with decreasing gravity.

A second example of the use of synthetic colours to study what is actually measured is a photometric investigation of the flux around 1.6 μ in late-type stars, started by Lennart Nordh and Göran Olofsson. They found an unexpected luminosity effect such that the 1.6 μ flux relative to the flux at shorter wavelengths was greater in giants than in supergiants. Roger Bell and I could quantitatively explain this as a result of increased blocking in the 1.6 μ band, caused by CN and CO lines (Bell, Gustafsson, Nordh and Olofsson 1976).

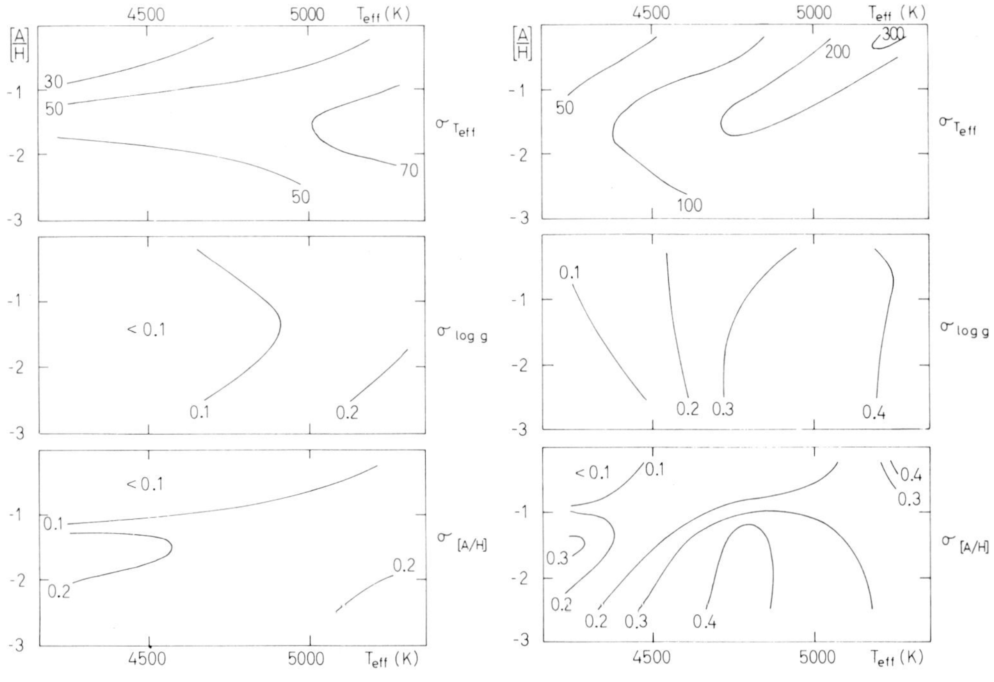

Figure 8. The errors introduced into the determinations of fundamental parameters (T_{eff}, log g and [A/H] from top to bottom) for red giants by errors of $0\overset{m}{.}01$ in the observations, when the DDO system is used. Interstellar reddening is assumed to be well known (left) or assumed to be unknown and to be determined from the observations (right). Log g was chosen to be 2.625 but the diagrams are similar for other gravities. (Figures from Paper I.)

We may also use the synthetic colours to obtain *measures of the power of photometric systems*. Once a grid of synthetic colours has been constructed it is a fairly simple task to derive diagrams like Fig. 8, showing the errors in the fundamental parameters, resulting from assumed errors of $0\overset{m}{.}01$ in the observed indices, if the DDO system is used for determining these parameters for red giants. These errors in the parameter estimates are dependent on T_{eff}, log g and [A/H] and are displayed in Fig. 8 for a given gravity as functions of T_{eff} and [A/H]. The interstellar reddening, E(B-V), was a) assumed to be known and b) assumed to be a free parameter, to be determined from the observations. Forming suitable means over the relevant T_{eff}, log g, [A/H] space one may condense this information to a table which makes it possible to rank the photometric systems with respect to their power (cf. Table 2). Note, however, that the bandwidths are widely different and that some systems contain more ultraviolet bands, which make them more expensive to use for late-type stars. There may also be other advantages, and disadvantages, of the systems which are not properly described by Table 2. It should also be noted that the data given in Table 2 are mainly relevant for investigations of Population II giants while the power of the systems in studies of Population I stars may be rather different. This is mainly due to the fact that some indices which are sensitive to metal abundance for metal-poor models tend to saturate for metal-rich ones.

One might also think of using the synthetic-colour technique to *construct new, still more efficient photometric systems*, e.g. with broader bands than those used in most narrow-band systems. Such work is underway but we have no specific results to report here.

Finally, one may use this technique for more special studies, e.g., *to study the effects on colours from interstellar or terrestrial extinction or to suggest transformation relations between different systems*. I shall show only one example of such an application of the technique - the terrestrial extinction coefficient in U-B varies with U-B in different ways for different metal abundances (cf. Fig. 9). Thus, if somebody determines the coefficient using solar metal-abundance stars and then observes a globular cluster at low altitudes (and many people do so) it will be dangerous to interpret the observed ultraviolet excesses as metal-abundance effects.

IV. HOW DO I GET MY COLOURS SYNTHESIZED?

Any photometric stellar observer, and especially those who establish new photometric systems, should be aware of the possibility that, sooner or later, there may develop an interest in attempting a theoretical calibration of the system. A very important requirement from the theoreticians will then be that the system is well documented, not only by some colours for some standard stars, but also that filter profiles have been measured and checked for stability during the

Table 2. Errors in the fundamental parameters, deduced for red giant stars from their colours, based on assumed observational errors of $0^{m}.01$ in each colour. The values are the medians of the standard deviations in the fundamental parameters in the T_{eff} - log g - [A/H] space covered by the models grid (adapted from Paper I).

System	*E (B − V) = 0.0*			*E (B − V) = free parameter*			
	$<\sigma_{Teff}>$ (K)	$<\sigma_{\log g}>$	$<\sigma_{A/H}>$	$<\sigma_{Teff}>$ (K)	$<\sigma_{\log g}>$	$<\sigma_{A/H}>$	$<\sigma_{E(B-V)}>$ (mag.)
UBVRI	47	0.72	0.39	495	2.79	1.47	0.15
Geneva	39	0.21	0.12	257	0.78	0.52	0.08
without U	84	1.26	0.30	261	1.33	0.55	0.08
uvby	57	0.15	0.13				
with R-I	37	0.13	0.11	226	0.56	0.58	0.06
uby with R-I	67	0.95	0.57				
DDO	47	0.10	0.13	185	0.48	0.30	0.09
without 35 and 41 bands	88	0.28	0.22				
with B-V	30	0.08	0.09	115	0.26	0.23	0.04
with B-V without 41 bands	42	0.09	0.12	185	0.32	0.38	0.06
with B-V without 35 and 41 bands	48	0.25	0.13	283	0.64	0.55	0.09
gnkmf	28	0.08	0.09	103	0.23	0.24	0.06
gkmf	57	0.10	0.17	398	0.65	0.86	0.16
Uppsala with B-V	44	0.30	0.18				

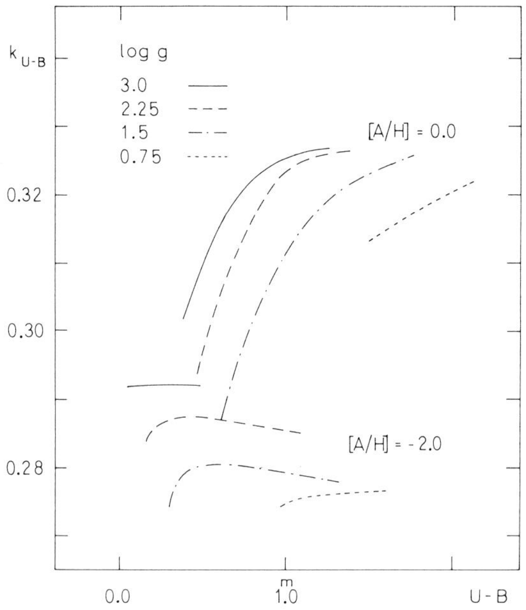

Figure 9. The terrestrial atmospheric extinction coefficient in U-B, k_{U-B}, for models with different values of U-B, log g and [A/H]. The interstellar extinction is neglected. Note the differences for models with different metal abundance. (Figure from Paper I.)

observing periods. Especially if the bands are broad, the transmission functions for the whole telescope-photometer system also ought to be measured. Much of the effort involved in computing synthetic colours has gone into a struggle to correct obviously erroneous sensitivity profiles given in the literature. Moreover, if the colours observed with such a carefully documented instrumentation are transformed to some other ("standard") system before being published, this fact, and the transformation relations, should be given. Also empirical terrestrial extinction coefficients are valuable data for the theorist - as a rough check on the accuracy of the filter profiles if nothing else.

A further requirement will be that some "normal" and well-observed stars, suitable as standards for determining the zero-points, have been observed in the photometric system. In studies of late-type stars, one should not only observe some early-type zero-point stars but also some well-studied red giants, like ϕ^2 Ori (cf. Bell and Gustafsson 1975) or G2 V dwarfs similar to the Sun.

Having done this the observer could approach a group producing model atmospheres and synthetic spectra for stars in the relevant spectral range. If he or she has good reasons for asking their assistance, I assume that they will often be helpful.

V. SHOULD WE *COMPUTE* COLOURS INSTEAD OF *OBSERVING* THEM?

This paper hopefully has demonstrated that the synthetic-colour technique can be used today for many purposes and will be used for still more in the future. But the best and probably the only way to determine the colour of a star is to observe it. The choice of how to measure and analyse the results may be inspired and improved by using synthetic colours. However, there are many fundamentals of stellar atmospheres that we do not understand properly today, and many more trivial data we do not know very well. Like any other human system based on oversimplifying ideas, optimistic extrapolations and complicated computations the synthetic colours should be treated with suspicion. Use them, but use them with care - and do not forget about the sometimes widely different and more interesting reality!

ACKNOWLEDGEMENTS

It is a pleasure to thank the pioneer within this field of research, Roger Bell, for a most interesting and stimulating collaboration. His active participation has been a necessary condition for all the results discussed above.

REFERENCES

Bell, R.A. and Gustafsson, B.: 1975, Dudley Obs. Report No. 9, *Multicolour Photometry and the Theoretical HR Diagram,* eds. A.G. Davis Philip and D.S. Hayes, p. 319.

Bell, R.A. and Gustafsson, B.: 1976, R. Greenwich Obs. Bull. No. 182, *The Galaxy and the Local Group,* eds. R.J. Dickens and J.E. Perry, p. 109.

Bell, R.A. and Gustafsson, B.: 1978, Astron. Astrophys. Suppl. Ser., in press.

Bell, R.A., Eriksson, K., Gustafsson, B. and Nordlund, Å.: 1976a, Astron. Astrophys. Suppl. Ser. 23, 37.

Bell, R.A., Gustafsson, B., Nordh, H.L. and Olofsson, S.G.: 1976b, Astron. Astrophys. 46, 391.

Bell, R.A., Dickens, R.J. and Gustafsson, B.: 1978, Astrophys. J., in press.

Buser, R. and Kurucz, R.L.: 1978, submitted to Astron. Astrophys.

Gascoigne, S.C.B., Norris, J., Bessel, M.S., Hyland, A.R. and Visvanathan, N. 1976, Astrophys. J. 209, 25.

Gustafsson, B. and Ardeberg, A.: 1978, in *Astronomical Papers dedicated to Bengt Strömgren,* eds. A. Reiz and T. Andersen, Copenhagen University Obs., p. 145.

Gustafsson, B. and Bell, R.A.: 1978, Astron. Astrophys., in press, *Paper I.*

Gustafsson, B., Bell, R.A., Eriksson, K., Nordlund, Å.: 1975, Astron. Astrophys. 42, 407.

Gustafsson, B., Bell, R.A., Hejlesen, P.M.: 1977, Astrophys. J., 216, L7.

Hartwick, F.D.A. and McClure, R.D.: 1974, Astrophys. J. 193, 321.

Hejlesen, P.M.: 1976, private communication.

Kurucz, R.L.: 1978, submitted to Astrophys. J. Suppl. Ser.

Lambert, D.L., Ries, L.M.: 1977, Astrophys. J. 217, 508.

Lub, J. and Pel, J.W.: 1977, Astron. Astrophys. 54, 137.

Mould, J.R.: 1978, Astrophys. J. 220, 434.

Nissen, P.E. and Gustafsson, B.: 1978, in *Astronomical Papers dedicated to Bengt Strömgren,* eds. A. Reiz and T. Andersen, Copenhagen University Obs., p. 43.

Peytremann, E.: 1975, Astron. Astrophys. 38, 417.

Relyea, L.J. and Kurucz, R.L.: 1978, Astrophys. J. Suppl. Ser., in press.

Zinn, R.: 1978, Astrophys. J. 225, 790.

MASS LOSS AND STELLAR EVOLUTION

A. Renzini
Osservatorio Astronomico, Bologna, Italy

ABSTRACT

The available observational evidence about stellar winds and mass loss is briefly reviewed, and its implications for stellar evolution are discussed. Particular emphasis is placed on post-main sequence evolution of low-mass and intermediate-mass stars, including the final evolution from red giant to the white dwarf stage. For massive stars, as well as for the implications of stellar mass loss for supernovae and supernova remnants, the reader is referred to other recent review papers.

It is shown that evolutionary models, including mass loss at a rate consistent with the empirical one, provide a more satisfactory account than do constant mass evolutionary models.

1. INTRODUCTION

For a long time the importance of mass loss for the evolution of stars has not been properly recognized. Some 25 years ago just the opposite process, i.e. accretion from the interstellar medium, was seriously considered. Even objects like planetary nebulae and the solar corona were frequently interpreted in terms of accretion! Only quite recently there has been a renewed interest in stellar mass loss, both under the pressure of direct observational evidence and in the attempt to explain a number of astrophysical topics which are not satisfactorily accounted for by constant-mass evolutionary models. It is worth emphasizing that the importance of stellar mass loss is not confined to the problems of stellar evolution, as the chemical evolution of galaxies is primarily controlled by the amount of material the stars restitute to the interstellar medium and by its composition.

Bengt E. Westerlund (ed.), Stars and Star Systems, 155–171.

Apart from the supernova explosion stars can lose mass by two different kinds of processes namely: i) by a (more or less) steady stellar wind, and ii) by pulsational and/or dynamical instabilities of the outer stellar envelope.

Section 2 of this paper briefly reviews the most relevant direct observational evidences for stellar mass loss, while section 3 is devoted to a discussion of the effects of mass loss on stellar evolution. Finally, section 4 provides a short review of the problems encountered in understanding the physical mechanisms driving stellar winds.

2. OBSERVATIONAL EVIDENCES FOR STELLAR WINDS

Observations clearly indicate the existence of a radial mass outflow for three classes of stars: i) OB and early A type stars of all luminosity classes, ii) late type giants and supergiants, and iii) the sun. Other stars, and possibly all, presumably support a stellar wind, but our diagnostic tools are inadequate to detect such winds when either the mass loss rate (MLR) is too small or the distance is too large. With our present observational techniques it is impossible to detect a solar like wind from a star similar to the sun and situated at a distance of a few parsec.

In general, we shall consider a wind as 'important' when either the MLR is large enough to decrease significantly the stellar mass during the evolution of the star (by, say, more than 5 - 10 percent), or when the mass outflow is able to affect the surface composition of the star, which eventually exposes nuclearly processed layers. The sun, unique representative in its class, has been mentioned in spite of its low MLR ($\sim 2\times10^{-14}$ $M_{\odot}$/yr) because solar studies can be of great help in understanding the physical mechanism(s) responsible for the wind in the other kind of objects.

2.1 Stellar winds in hot stars

In early type stars the presence of a wind is revealed by the P-Cygni type line profile exhibited by several spectral lines, particularly impressive being those in the ultraviolet. There is no doubt that the circumstellar material is actually leaving the star, since the flow terminal velocities, inferred fitting the P-Cygni line profiles, are always larger than the escape velocity from the stellar surface. The existence of a wind is also inferred from the infrared excess due to free-free emission in the circumstellar envelope (CSE). For this kind of stars Lamers and Morton (1976), Lamers and Rogerson (1978) and Barlow and Cohen (1977) among others, have determined MLRs ranging from $\sim 2\times10^{-10}$ $M_{\odot}$/yr up to $\sim 10^{-5}$ $M_{\odot}$/yr. Lamers *et al* (1976) have obtained an analytical fit of the observations relating the MLR to the basic stellar parameters: luminosity, mass and radius. In general, observed MLRs increase for increasing luminosity and effective temperature, the largest MLRs being found in most luminous O type supergiants. Terminal wind velocities are typically of a few thousand km/s.

Besides normal hot main-sequence and post main-sequence stars also some highly evolved hot stars appear to support a substantial wind. Large MLRs (typically around 10^{-5} $M_\odot$/yr) are inferred for WR stars (cf. Smith 1974), while the WR or Of appearance of the spectrum of several planetary nebula nuclei, PNN (Smith and Aller 1969), indicates that presumably also PNN are supporting a stellar wind. However, the author is not aware of any MLR determination for PNN stars (which would be of great value).

2.2 Stellar winds in red giants and supergiants

The presence of a stellar wind around late type giants and supergiants is revealed by several observational evidences:
i) narrow circumstellar absorption and/or P-Cygni low excitation lines of neutral or singly ionized metals (Ca, Na, K, Ba, etc.) - among the most recent works, see Reimers (1975a, b; 1977a), Sanner (1976), Bernat (1977) and references therein. Terminal wind velocities are always smaller than the escape velocity from the surface, but they exceed the local escape velocity as revealed by the CS absorption lines in the spectra of near visual companions of red giants (typical the case of α Her; Reimers 1977b).
ii) Chromospheric emission (CaII H& K, MgII h & k), cf. Bernat and Lambert (1976a) and references therein.
iii) Circumstellar Hα emission in population II red giants (Cohen 1976; Mallia and Pagel 1978)
iv) Direct resolution of the extended CSE (Bernat and Lambert 1976b)
v) Circumstellar dust emission around 11μ and 18-20μ (Humphreys 1974, and references therein)
vi) Molecular maser lines in the microwave region, in particular OH lines at 1612 MHz (Wilson and Barrett 1972; Habing 1977, and references therein; Olnon 1977).

The first method (observations of CS absorption lines) provides the most reliable MLRs. Reimers (1975a,b) has given an analytical fit to the observations, relating the MLR to the basic stellar parameters:

$$\dot{M} = -\ 4\text{x}10^{-13}\ \eta\ L/gR \qquad (M_\odot/\text{yr}) \tag{1}$$

where L,g (the surface gravity) and R are in solar units and η (an uncertainty factor) is between 1/3 and 3. A recalibration of this MLR expression (MLRE) gives $\eta = 0.35 \pm 0.2$ (Reimers 1977 a,b). However, this estimate has been further revised, restoring a value close to the original one (Reimers 1978; Kudritzki and Reimers 1978). Therefore a prudent estimate still gives $1/3 < \eta < 3$, on purely observational grounds. Observed MLRs range from $\sim 10^{-9}$ $M_\odot$/yr for K giants up to $\sim 10^{-5}$ $M_\odot$/yr for M supergiants. The MLR appears to decrease rapidly for increasing stellar effective temperature, at constant luminosity, while it seems independent of metal abundance (Reimers 1975 a,b).

Terminal wind velocities range from ∿10 km/s in M supergiants to ∿75 km/s in K giants.

2.3 Planetary nebulae

Planetary nebulae, PN, represent one of the most obvious indications that stars lose a significant amount of mass during their advanced evolutionary phases. Unfortunately, the determination of the nebular mass is quite uncertain, as also the distance must be determined (cf. O'Dell 1962; Cahn and Kaler 1971; Perinotto 1975). Using all available data, Perinotto (1975) finds $<M_N/M_\odot> = 0.15$ for the average nebular mass, but $< \log M_N/M_\odot > = -1.5$ corresponding to $M_N = 0.03\ M_\odot$. The spread in M_N values is presumably rather large, owing to the wide mass spectrum of precursor stars (cf. section 3.3). Planetary nebulae are generally believed to originate from the sudden (or fast) envelope ejection due to a dynamical envelope instability in low- and intermediate-mass stars, during their asymptotic giant branch (AGB) phase. Multiple shell structure of some PN (cf. Kaler 1974) may indicate that such envelope instability can occur more than once during the final AGB evolution, being possibly triggered by the last few relaxation cycles following the corresponding helium-shell flashes. The observation of extended, low surface brightness halos in several PN (Millikan 1974) has been interpreted as evidence for the red giant wind preceding envelope ejection (Fusi-Pecci and Renzini 1976).

The expansion velocity of PN is typically around 20 km /s, and its composition, reflecting the composition of the stellar envelope just prior to the ejection, is an important source of information about the initial mass, M_i, of the parent star and its evolutionary history.

3. STELLAR EVOLUTION WITH MASS LOSS

Only quite recently, has mass loss been incorporated in stellar evolutionary calculations, even if mass loss has been frequently invoked to appease a number of discrepancies between observations and the results of stellar evolution theory. Mass loss being a surface phenomenon one can reasonably expect that MLRs depend primarily on some combination of basic stellar parameters rather than on the interior structure of the star. For instance, this is the case for Eq. (1) and for the MLR expression given by Lamers *et al* (1976). Therefore, in discussing stellar mass loss an important question of method is to consider all types of stars (i.e. all initial masses and all evolutionary phases), applying the available MLR expressions in the pertinent evolutionary phases and determining the efficiency parameter(s) (like η in Eq. (1)) by requiring the evolutionary tracks to fulfil the observational constraints. In such a way, i) justified claims for mass loss can be distinguished from purely *ad hoc* ones, ii) MLRs can be more precisely estimated than from direct observations, iii) some insight can be obtained on the physics itself of the mass loss process. In fact, both observations and theory of mass loss are presently rather uncertain, while important details of stellar evolution crucially depends on the assumed MLR. In other words, an astrophysical approach is now more rewarding than a purely physical one. For these reasons, in the following, low mass as well as intermediate mass and massive stars will be considered.

3.1 Low mass stars ($M_i \sim M_\odot$)

For low mass stars the most interesting astrophysical case concerns population II. It is known since long that constant mass evolutionary models fail to adequately reproduce the horizontal branch (HB) morphologies of globular clusters (cf. Castellani and Renzini 1968; Iben and Rood 1970; Rood 1973). Without mass loss, theoretical HBs are much redder than the observed ones, and a mass loss of about 0.2 $M_\odot$ (prior to the HB phase) is required to remove this discrepancy. Some dispersion about this value ($\sim$ 0.025 $M_\odot$; Rood 1973) is also required for stars within one cluster in order to reproduce the observed HB extension. Further, during the post-HB phase (i.e. the AGB phase), in absence of mass loss along the AGB, globular cluster stars would reach too high luminosities, which also conflicts with the observations (cf. Renzini 1976, 1977). To prevent this contradiction, $\sim$ 0.1 $M_\odot$ must be lost during the AGB phase. Therefore, substantial mass loss must occur along both giant branches, in such a way that a typical population II star with initial mass $M_i \approx 0.8$ - 0.85 $M_\odot$ will eventually die as a white dwarf of final mass $M_f \approx 0.5$ - 0.55 $M_\odot$. Mass loss by stellar wind in population II stars ascending the red giant branch (RGB) and the AGB has been discussed by Fusi-Pecci and Renzini (1975 a,b; 1976; 1977) and Renzini (1976, 1977). The main result is that the above requirements are both fulfilled using Eq. (1) in conjunction with available evolutionary rates, if $\eta = 0.38 \pm 0.02$. Larger values of η would give rise to exceedingly blue HBs, while for $\eta \gtrsim 0.5$ the HB phase would completely disappear, the whole H-rich envelope being lost prior to helium ignition in the core.

AGB evolution of population II stars is probably terminated by the ejection of the remaining envelope, as indicated by the PN Observed in M15, whose mass is $\sim$ 0.018 $M_\odot$ (Peimbert 1973). Therefore, observed red giant MLRs perfectly account for the HB and AGB morphology of globular clusters (cf. also Cohen 1976 ; Mallia and Pagel 1978). Furthermore, the extreme sensitivity of the HB morphology to η allows an extremely precise determination of this "efficiency parameter" in population II red giants (an uncertainty of only 4% is much smaller than the observational one). Globular cluster stars provide also another important information on red giant winds, namely, on the dependence of the MLR on metal abundance. It is well known that globular clusters exhibit a wide range of metal abundances, from $Z \sim 10^{-4}$ to $\sim 10^{-2}$ (which is almost solar). Metal poor clusters have typically a blue HB, while metal rich ones have red HBs. Assuming $\dot{M} \propto Z^{\alpha}$ for the red giant MLR, this fact indicates that $\alpha < \sim 0.2$, otherwise stars in metal rich clusters would lose too much mass producing blue HBs (which is contrary to the observations). Observed HBs are easily reproduced with $\alpha \simeq 0$. This implies that the MLR in population II red giants is *almost independent of metal abundance,* a point which must be seriously considered in discussing the physical mechanism producing the red giant wind.

Concerning population I low mass stars, Dearborn *et al.* (1976) and Dearborn and Eggleton (1976) have recently invoked substantial mass loss (up to half the stellar mass) during the main-sequence or sub-giant phases in order to account for the low $^{12}C/^{13}C$ ratio apparently observed in several K giants. Substantial mass loss prior to the deep penetration of the outer convection zone - and the corresponding dredging up of ^{13}C - would, in fact, reduce the envelope mass over which ^{13}C is diluted, and lower $^{12}C/^{13}C$ ratios would result. However, there is no evidence for large MLRs during either the main-sequence or sub-giant phases of low mass stars. According to Eq. (1) most mass loss occurs close to the red giant tip (cf. Fusi-Pecci and Renzini 1975a), while a negligible amount of mass is lost during the sub-giant phase. The opinion of the writer is that the cause of the low $^{12}C/^{13}C$ ratios should be searched for in something else than mass loss (cf. Sweigart and Mengel 1978). Data reduction would also deserve a careful discussion owing to the large difference in the saturation effects between molecular bands with ^{12}C and those with ^{13}C. An underestimate of the ^{12}C saturation would in fact lead to an underestimate of the $^{12}C/^{13}C$ ratio.

Mass loss during the main-sequence and/or sub-giant phases of low mass stars has also been suggested as a possible explanation of some apparent peculiarities of the old galactic cluster NGC 188 (McClure and Twarog 1977; Twarog 1978). A first argument is based upon the apparently more dispersed distribution of giants with respect to main-sequence stars. The statistical evidence is however marginal, and further membership studies would be useful. The second argument concerns the different location in the HR diagram of the RGB of NGC 188 with respect to the RGB of M67. However, this difference could also arise from other causes owing to extant uncertainties in differential reddening, metallicity and distance between the two clusters.

In conclusion, low mass stars lose an important fraction of their mass on the RGB and on AGB, when they are in the hydrogen-shell and double-shell burning phases, respectively. Total mass loss along the RGB rapidly decreases for increasing initial stellar mass, becoming negligible for $M_i \gtrsim 1.5\ M_\odot$ (cf. Fusi-Pecci and Renzini 1976), while the opposite trend is exhibited by the mass loss along the AGB. No compelling evidence exists for appreciable mass loss during the pre-RGB or HB phases.

3.2 Intermediate mass stars ($M_i \lesssim 8\ M_\odot$)

The upper initial mass limit for "Intermediate mass stars" is defined as the maximum initial mass for which a degenerate carbon/oxygen core is developed during the double shell burning phase (df. Paczynski 1971). The value of 8 $M_\odot$ has been recently questioned, being possibly as low as 5 or 6 $M_\odot$ (see Renzini 1978, for a review). The lower mass limit is vague, as at least part of intermediate mass stars has an evolutionary history similar to low mass stars, eventually leaving a WD remnant. For these stars, according to Eq. (1), very little mass is lost prior to helium ignition in the core. Extrapolating the Lamers *et al.* (1976) MLR for hot stars to lower luminosities and effective temperatures, negligible mass loss is also predicted during the core hydrogen-burning phase (main-sequence phase). Therefore, these stars initiate the core helium-burning phase with a mass essentially equal to the original one. This result conflicts with a significant mass loss prior to the cepheid phase, frequently claimed to account for an apparent discrepancy between the cepheid mass obtained using evolutionary models and pulsation theory (cf. Cox *et al.* 1977, and references therein). Estimated "pulsational" masses are ∿ 50 % (and even more) lower than "evolutionary" ones, this is the so called "Cepheid mass discrepancy". Mass loss during the core helium-burning phase is also inadequate to account for the discrepancy. According to Reimers (1975a,b) Eq. (1) provides an upper limit for the MLR in stars bluer than the RGB. An even more generous upper limit for the MLR in core helium-burning stars is obtained using Eq. (1) as transformed in Fusi-Pecci and Renzini (1976):

$$\dot{M} = -1.4\times10^{-13}\,\eta\, M^{-1.16} L^{1.68} \qquad (2)$$

which pertains to stars lying on the Hayashi track and with Z = 0.02. Such upper limits are listed in Table I for various initial masses and $\eta = 1$ and used to predict an upper limit for the total mass loss $-\Delta M$ - during the whole core helium-burning phase for stars with $M_i \leq 12.6\ M_\odot$. Average luminosities and core helium-burning lifetimes are from Hejlesen (1977).

Table I

$M_i/M_\odot$	$\log L/L_\odot$	Δt (10^6 yr.s)	$\dot{M}$ ($M_\odot$/yr)	ΔM ($M_\odot$)
12.6	4.58	1.92	3.6×10^{-7}	0.69
7.9	3.90	5.60	4.5×10^{-8}	0.25
5.0	3.10	21.50	3.5×10^{-9}	0.08
3.2	2.30	100.0	2.6×10^{-10}	0.026
2.5	1.90	188.0	7.4×10^{-11}	0.014

This result agrees with the arguments of Iben and Tuggle (1972, 1975) according to which revised distance and/or temperature scales, rather than mass loss, can easily explain the "cepheid mass discrepancy" (see also Pel 1976); Fusi-Pecci and Renzini 1976; Renzini 1977). Insisting on a mass loss $\Delta M \sim 0.5\ M_i$ one would derive η values with disastrous implications for low mass stars (cf. previous sub-section) and for the subsequent evolutionary phases of intermediate mass stars.

However, the mass discrepancy for double mode and bump cepheids cannot be eliminated by a simple revision of the distance and/or temperature scales. For these cepheids some other reason has to be found. Cox *et al* (1978) find that helium enhancement (Y = 0.75) in the very external layers of these stars could account for the observed period ratios and bump phase with pulsational masses in agreement with evolutionary ones. Cox *et al* suggest that a moderate "cepheid wind" ($\dot{M} \sim 10^{-10} - 10^{-11}\ M_\odot$/yr, cf. with Table I) but substantially helium deprived compared to the photospheric and/or envelope abundance, would produce such high helium abundance in the external layers. A helium-poor wind is not such an *ad hoc* assumption as at first sight it could seem. The solar wind, the best known stellar wind, is in fact helium-poor. However, if this characteristic is shared also by RGB stars, sizable helium enhancements could be produced in the whole envelope of population II red giants, affecting the instability strip location during the subsequent HB phase and the HB lifetime itself. If this would be the case the helium abundance determination of globular clusters would also be affected, while available determinations give Y-values in substantial agreement with the big-bang estimates (cf. Renzini 1977, and references therein). In conclusion, no completely satisfactory explanation of the properties of the double-mode and bump cepheids is presently available. The subsequent evolution (i.e. double shell burning phase) of intermediate mass stars is one of the most complex stages encountered in stellar evolution. This is so because of the uncommon variety of physical processes going on in AGB stars, namely: i) helium-shell flashes and subsequent thermal relaxation cycles, ii) convective envelope penetration and dredgning up of helium-burning and s-process products (Iben 1976; Iben and Truran 1978), probably originating carbon stars, iii) substantial stellar wind, producing massive CSEs (dusty CSEs, stellar masers, etc.), iv) pulsational envelope instability (Mira variables and other red variables), v) dynamical envelope instability causing envelope ejection and the evolution towards the PNN and WD stages (for $M_i < M_{crit}$) or vi) carbon ignition in the degenerate carbon/oxygen core ($M_i > M_{crit}$) probably followed by a SN outburst. The mutual interaction of some of these processes could also give rise to even more complex physical situations. Of course, all these topics cannot be properly discussed in this short review and I will concentrate on a few points, only. Mass loss by stellar wind during the AGB phase has been discussed by Fusi-Pecci and Renzini (1975b, 1976), Renzini (1976, 1977), Mengel (1976), Scalo (1976), Wood and Cahn (1977), and Weidemann (1977 a). The main point is the determination of M_{crit}, which, according to a number of astrophysical indications, should be in the range between 4 and 6 $M_\odot$ (cf. Tinsley 1977; Wheeler 1978). For $M_i < M_{crit}$ wind mass loss and the subsequent dynamical envelope ejection prevent the degenerate C-O core from reaching the Chandrasekhar limit ($\sim 1.4\ M_\odot$), and the star can die as a white dwarf. When only the effect of the wind is taken into account, one finds $M_{crit} = 4\ M_\odot$ for $\eta = 0.58$ (Fusi-Pecci and Renzini 1976) and $M_{crit} = 6\ M_\odot$ for $\eta = 2$. Uncertainties in the red giant MLR do not allow a precise determination of M_{crit}, but there is no doubt that the red giant wind plays a fundamental role. As far as the dynamical envelope instability is concerned extant hydrodynamical studies (Kutter and Sparks 1974; Stry 1975, and references therein) do not presently allow us to know the mass of the ejected envelope as a function of M_i. Further hydrodynamical calculations would be very valuable. In this respect the semiempirical approach of Wood and Cahn (1977) is very instructive. Taking into account both types of mass loss they are able to reproduce the observed frequency-period histogram for Mira variables with $0.25 < \eta < 0.5$, implying $3.7 < M_{crit} < 4.7$. Values of η outside this range would give either too many or to few Miras and period distributions at variance with the observations. For a review on the problem of the determination of M_{crit} and, in general, on the importance of stellar winds for supernovae and SN remnants see Renzini (1978).

3.3 Evolution from the AGB to the WD stage

Stars with $M_i < M_{crit}$ continue to ascend along the AGB undergoing helium-shell flashes and thermal relaxation cycles, until the mass of the H-rich envelope, M_e, falls below a critical value, $M_{e,min}$. When $M_e < M_{e,min}$ the star initiates a blueward excursion, crosses the PNN region, and eventually fades and cools approaching the WD cooling line. Three processes cooperate in reducing M_e below the critical value: i) The increase in the core mass, M_c, at the expense of M_e through the shell hydrogen burning, ii) stellar wind, and iii) fast envelope ejection due to the envelope instability. In the case of $M_c = 0.6\ M_\odot$ Gingold (1974) finds $M_{e,min} = 0.005\ M_\odot$, a value very close to the amount of the core mass increase, ΔM_c, during the previous relaxation cycle ($0.007\ M_\odot$ in this case). Although there are no other similar models available, we conjecture that $M_{e,min}$ is always close to ΔM_c irrespective of the value of M_c. Since ΔM_c as a function of M_c can easily be obtained from the relations given by Paczynski (1970, 1971), we would have:

$$M_{e,min} \gtrsim \Delta M_c \approx 1.9\text{x}10^{-4}\ (5.95\ M_c - 3.1)\ 10^{-4.5\ (M_c - 1,0)} \qquad (3)$$

or $M_{e,min}$ ranging from $5\text{x}10^{-3}$ to $10^{-5}\ M_\odot$ for M_c between 0.6 and and 1.4 $M_\odot$. In other words, we conjecture that stars leave the AGB when the envelope mass becomes insufficient to provide H-rich material for completing a further quiescent shell hydrogen-burning phase.

The internal structure of stars leaving the AGB is therefore the following: i) a degenerate carbon-oxygen core containing almost all the stellar mass, ii) a non-degenerate heliumrich intershell region of mass ΔM_{He} given by:

$$\log \Delta M_{He} \simeq -2.75 - 2.8(M_c - 1) \qquad \text{for } M_c < 1$$

$$\simeq -2.75 - 6.0(M_c - 1) \qquad \text{for } M_c > 1 \qquad (4)$$

(Paczynski 1975), and iii) a H-rich envelope of mass $M_e < M_{e,min}$. Two zones can be distinguished in the intershell region, the outer zone being (at most) of similar mass compared to the inner one. The outer zone contains products of CNO H-burning, i.e. helium and nitrogen, while the inner one is composed of helium and helium-burning products, since it has been mixed by the convective shell during the previous shell-flash. Its carbon abundance is therefore:

$$X_C \simeq 0.25\ (M_c - 0.16) \qquad (5)$$

(Iben and Truran 1978). Only a few post-AGB models, consistent with the previous prescriptions, have been computed (e.g. Gingold 1974; Paczynski 1971). As far as the comparison with the observation of PNN is concerned Paczynski (1970, 1971) noted a good *qualitative* agreement. However, there is a striking *quantitative* disagreement between Paczynski's evolutionary lines and the location in the HR diagram of PNN according to O'Dell (1968) or according to the so-called Harman-Seaton (H-S) sequence (Harman and Seaton 1966). It is true that evolutionary lines "go through" the O'Dell or H-S locus, but the brightest 'observed' points would require a mass of the post-AGB star close to 1.4 $M_\odot$ (or even larger!). Such massive post-AGB stars spend a very short time at high luminosities, for instance the 1.2 $M_\odot$ Paczinski's model takes only 5.6 years to move from the AGB to the maximum temperature reached along the post-AGB track! This time would be even shorter for more massive models. Observed PN have diameters between 0.08 and 1.4 pc, and expansion velocities of $\sim$ 20 km/s. Therefore PNN have ages between $\sim$ 2000 and $\sim$ 34000 years, starting from envelope ejection and the subsequent departure from the AGB. Younger PN are not observed either because the post-AGB star is not yet hot enough to excite the nebula, or because the time is becoming too short and the corresponding discovery probability too small. Older PN are not observed either because the central star is correspondingly too faint, or, more likely, because the nebula has already dispersed. Fig. 1 shows the Paczynski evolutionary lines for M = 0.6, 0.8 and 1.2 $M_\odot$. The time required to reach the numbered points along the tracks can be found in Paczynski (1971). Point A on the tracks corresponds to t = 2000 yrs and point B to t = 34000 yrs, while the boxes enclose the part of the track between t = 3800 and 34000 yrs. The boxes for the 0.7 and 0.55 cases have been guessed by interpolation or extrapolation. Therefore, the boxes define a rather wide band in the HR diagram, where PNN are expected to be found. The "average" H-S

computed so far. Therefore, the possible effects of mass loss on post-AGB stars can only be guessed on the basis of existing information. Mass loss during the red phase, when the star is still close to AGB, will probably have just the effect of accelerating the blueward movement of the star in the H-R diagram (cf. Härm and Schwarzschild 1975). It is worth emphasizing that mass loss cannot significantly affect the total mass of the star since the time spent in this phase is very short. However, we are dealing with very low mass H-rich envelopes and M_e continues to decrease as the star evolves, since during the post-AGB phase the major energy source is still H-shell burning. According to Paczynski (1971), in the last computed models for M = 0.6, 08, and 1.2 $M_\odot$ M_e was reduced to 1.4×10^{-4}, 2.0×10^{-5} and 7.5×10^{-7}, respectively. Even with relatively modest MLRs ($\sim 10^{-9} - 10^{-7}$ $M_\odot$/yr) mass loss is likely to compete with H-burning in reducing M_e, and therefore in determining the evolution of post-AGB stars. It is worth noting that low mass stars will be more likely affected by mass loss as they spend a longer time at high luminosities (when presumably MLRs are large). Models including mass should be computed in the future in order to achieve quantitative results.

Furthermore, mass loss during the post-AGB phase opens a very fascinating perspective, namely the complete removal of the residual H-rich envelope. Very little mass ($10^{-4} - 10^{-3}$ $M_\odot$) needs in fact to be lost in order to expose the intershell H_e -rich regions. It is very encouraging that in the (log T_{eff} - log g) -diagram most H-poor (or H-deprived) stars are found along the path followed by the post-AGB stars from the AGB to the WD region. These H-poor stars include: i) intermediate and extreme He-stars, ii) H-deficient PNN (WR-type and others), iii) H-deficient carbon stars (HdC) and R Coronae Borealis stars (R CrB), iv) H-deficient O sub-dwarfs, and, finally, v) non-DA white dwarfs. The possibility that all these objects are post-AGB stars, representing successive developments of a same event, needs to be carefully investigated by means of post-AGB models including mass loss. The fact that several Helium-stars, HdC and R CrB stars have an unusually high carbon abundance ($X_C \sim 0.1$) is a very strong indication that these stars are actually exposing the inner zone of the intershell region (cf. Eq. (5))! As far as we know, only in post-AGB stars do layers of this composition get so close to the surface. An important question is: during which stage of the post AGB phase can the H-rich envelope be completely removed? It does not appear likely that complete envelope removal occurs during quiescent shell H-burning since the effect of mass loss would be to quench the H-shell, driving the star to very low luminosities (and negligible MLRs). Conversely, it appears much more likely that envelope removal occurs when the H-shell is inactive, i.e. when the star is responding to the very last helium-shell flash. To this last flash are due to the wiggles on the evolutionary lines in Fig. 1. The time spent during such loops is a very sensitive function of the stellar mass: being $\sim$ 2000, $\sim$ 170, and only $\sim$ 20 years, respectively, for M = 0.6, 0,8 and 1.2 $M_\odot$. If the above speculations are indeed correct the mentioned bright H-poor stars would be in a loop phase, i.e. in the thermal relaxation cycle triggered by the final shell flash. As a consequence the present mass of such stars should be $M \lesssim 0.6\ M_\odot$, otherwise the relaxation cycle is too fast.

H -poor stars have been frequently interpreted in terms of the evolution of initially homogeneous H-poor or pure helium stars. How such curious stars originate has never been convincingly explained. An attempt to account for H-poor stars in terms of post-AGB models has been recently made by Schönberner (1977). However, his starting models do not have a self-consistent post-AGB structure, the mass of the helium envelope being much larger than the mass of the intershell zone given by Eq. (4). This is also the reason for which he finds quiescent helium-shell burning and red giant helium-star models, which would not be the case for a proper choice of ΔM_{He} according to Eq. (4). R CrB stars cannot immediately derive from AGB stars and cannot be in thermal equilibrium. According to the previous considerations they have *returned* to the red giant region, coming from the PNN sequence, in response to the final shell flash. Finally, it is worth mentioning that a fast wind from the PNN stars could have important effects on the structure and evolution of the PN themselves (cf. Kwok *et al* 1978).

3.4 Massive stars ($M_i \gtrsim 8\ M_\odot$)

The effects of mass loss during the main-sequence and the advanced evolutionary phases of massive stars have been recently reviewed by Chiosi (1977, 1978) and Renzini (1978).

4. THE QUESTION OF THE PHYSICAL PROCESS(ES) DRIVING-STELLAR WINDS

In spite of the growing body of observational information accumulated in recent years, the physics of stellar winds still awaits a satisfactory understanding. This holds for hot stars as well as for red giants and supergiants. Furthermore, the situation is only marginally better for the solar wind. Basically, two (conflicting) physical processes have been proposed both for hot and cool stars: i) a coronal-like expansion, as for the solar wind, and ii) radiation pressure due to ion lines (for hot stars) or on grains (for cool giants). According to the first hypothesis radiation pressure (on ions or on grains) could be important for a further acceleration of the flow, not to initiate it. As far as winds from hot stars are concerned, see Lamers and Rogerson (1978, and references therein), and I.A.U. Symposium No. 83 which has been entirely devoted to this problem. For cool giants see the reviews by Renzini (1977) and Weymann (1977).

Let us just mention here that the astrophysical considerations developed in sections 3.1 and 3.2 (implying that red giant MLR is almost independent of metal abundance) provide strong support to a coronal-like wind rather than to the grain-mechanism. Recent observations by Hagen (1978), indicating that the MLR is almost independent of the gas to dust ratio, provide further evidence disfavouring the grain-mechanism.

ACKNOWLEDGEMENTS

Drs F. D'Antona, F. Fusi-Pecci, Icko Iben Jr., I. Mazzitelli and M. Voli are gratefully acknowledged for stimulating discussions. This research was supported by grant 11.10.5561 of the Italian Ministry of Public Education.

REFERENCES

Barlow, M.J., Cohen, M.: 1977, Astrophys. J. 213, 737.
Bernat, A.P.: 1977, Astrophys. J. 213, 756.
Bernat, A.P., Lambert. D.C.: 1976a, Astrophys. J. 204, 830.
Bernat, A.P., Lambert, D.C.: 1976b, Astrophys. J. 210, 395.
Cahn, J.H., Kaler, J.B.: 1971, Astrophys. J. Suppl. Ser. 22, 319.
Castellani, V., Renzini, A.:1968, Astrophys. Space Sci. 2, 310.
Chiosi, C.: 1977, I.A.U. Symp. No. 80, in press.
Chiosi, C.: 1978, I.A.U. Symp. No. 83, in press.
Cohen, J.G.: 1976, Astrophys. J. 203, L127.
Cox, A.N., Deupree, R.G., King, D.S.,Hodson, S.W.: 1977, Astrophys. J. 214, L127.
Cox, A.N., Michaud, G., Hodson, S.W.: 1978, Astrophys. J. 222, 621.
Dearborn, D.S.P., Eggleton, P.P.: 1976, Q.J.R. Astron. Soc. 17, 448.
Dearborn, D.S.P., Kozlowski, M., Schramm, D.N.: 1976, Nature 261, 210.
Fuci-Pecci, F., Renzini, A.: 1975a, Astron. Astrophys. 39, 413.
Fusi-Pecci, F., Renzini, A.: 1975b, Mém. Soc. Roy. Sci. Liège 6e, Ser., 8, 383.
Fusi-Pecci, F., Renzini, A.: 1976, Astron. Astrophys. 46, 447.
Fusi-Pecci, F., Renzini, A.: 1977, I.A.U. Symp. No. 80, in press.
Fusi-Pecci, F., Renzini, A., Voli, M.: 1978, in preparation.
Gingold, R.A.: 1974, Astrophys. J. 193, 177.
Habing, H.J.: 1977, Veröff. Remeis-Sternwarte Bamberg, 11, Nr. 121, 401.
Hagen, W.: 1978, Astrophys. J. 222, L37.
Härm, R., Schwarzschild, M.: 1975, Astrophys. J. 200, 324.
Harman, R.J., Seaton, M.T.: 1966, Mon. Not. R. Astron.Soc. 132, 15.
Hejlesen, P.M.: 1977, I.A.U. Symp. No. 80, in press.
Humphreys, R.M.: 1974, Astrophys. J. 188, 75.
Iben, I. Jr.: 1976, Astrophys. J. 208, 165.
Iben, I. Jr., Rood, R.T.: 1970, Astrophys. J. 161, 583.
Iben, I. Jr., Truran, J.W.: 1978, Astrophys. J. 220, 980.
Iben. I. Jr., Tuggle, R.S.: 1972, Astrophys. J. 178, 135.
Iben. I. Jr., Tuggle, R.S.: 1975, Astrophys. J. 197, 39.
Kaler, J.B.: 1974, Astron. J. 79, 594.
Kaler. J.B., Iben. I. Jr., Becker, S.A.: 1978, Astrophys. J. 224, L63.
Kudritzki, R.P., Reimers. D.: 1978, Astron. Astrophys., in press.
Kutter, G.S., Sparks, W.M.: 1974, Astrophys. J. 192, 447.
Kwok, S., Purton, C.R., Fitzgerald, P.M.: 1978, Astrophys. J. 219, L125.
Lamers, H.J.G.L.M., Morton, D.C.: 1976, Astrophys. J. Suppl. Ser. 32, 715.
Lamers, H.J.G.L.M., Rogerson, J.B.: 1978, Astron. Astrophys. 66, 417.
Lamers, H.J.G.L.M., van den Heuvel, E.P.J., Petterson, J.A.: 1976, Astron. Astrophys. 49, 327.
Mallia, E.A., Pagel, B.E.J.: 1978, preprint.
McClure, R.D., Twarog. B.A.: 1977, Astrophys. J. 214, 111.
Mengel, J.G.: 1976, Astron. Astrophys. 48, 33.
Millikan, A.G.: 1974, Astron. J. 79, 594.
O'Dell, C.R.: 1962, Astrophys. J. 135, 371.
O'Dell, C.R.: 1968, I.A.U. Symp. No. 34, p. 369. Eds. D.E. Osterbrock, C.R. O'Dell, Reidel Publ. Co., Dordrecht.

Olnon, F.M.: 1977, Veröff. Remeis-Sternwarte Bamberg 11, No. 121, 596.
Paczynski, B.: 1970, Acta Astron. 20, 47.
Paczynski, B.: 1971a, Acta Astron. 21, 271.
Paczynski, B.: 1971b, Acta Astron. 21, 417.
Paczynski, B.: 1975, Astrophys. J., 202, 558.
Peimbert, M.: 1973, Mém. Soc. Roy. Sci. Liège 6e Ser., 5, 307.
Pel, J.W.: 1976, Astron. Astrophys. Suppl. Ser. 24, 413.
Perinotto, M.: 1975, Astron. Astrophys. 39, 383.
Pottasch, S.R., Wesselius, P.R., Wu, C.-C., Fieten, H., van Duinen, R.J.: 1978, Astron. Astrophys. 62 95.
Reimers, D.: 1975a, Mém. Soc. Roy. Sci. Liège 6e Ser., 8, 369.
Reimers, D.: 1975b, in *Problems in Stellar Atmospheres and Envelopes*, p. 229, Eds B. Baschek, W.H. Kegel, G. Traving, Springer Verlag, Berlin.
Reimers, D.: 1977a, Veröff. Remeis-Sternwarte Bamberg 11, Nr. 121, 559.
Reimers, D.: 1977b, Astron. Astrophys. 61, 217.
Reimers, D.: 1978, Astron. Astrophys. 67, 161.
Renzini, A.: 1976, R. Greenwich Obs. Bull. No. 182, 87.
Renzini, A.: 1977, in *Advanced Stages in Stellar Evolution*, p. 151 Eds. P. Bouvier, A. Maeder, Geneva.
Renzini, A.: 1978, *Supernovae and Supernova Remnants*, Mem. Soc. Astron. Italiana, in press.
Rood, R.T.: 1973, Astrophys. J. 184, 815.
Sanner, F.: 1976, Astrophys. J. Suppl. Ser. 32, 115.
Scalo, J.M.: 1976, Astrophys. J. 206, 215.
Schönberner. D.: 1977, Astron. Astrophys. 57, 437.
Smith, L.F.: 1974, Mem. Soc. Astron. Italiana 45, 367.
Smith, L.F., Aller, L.H.: 1969, Astrophys. J. 157, 1245.
Stry, P.E.: 1975, Astrophys. J. 196, 559.
Sweigart, A.V., Mengel, J.G.: 1978, preprint.
Tinsley, B.M.: 1977, *Supernovae*, p. 117, Ed. D.N. Schramm, Reidel Publ. Co., Dordrecht.
Twarog, B.A.: 1978, Astrophys. J. 220, 890.
Weidemann, V.: 1977a, Astron. Astrophys. 59, 411.
Weidemann, V.: 1977b, Astron. Astrophys. 61, L27.
Weymann. R.J.: 1977, Veröff. Remeis-Sternwarte Bamberg 11, Nr. 121, 557.
Wheeler, J.C.: 1978, *Supernovae and Supernova Remnants*, Mem. Soc. Astron. Italiana, in press.
Wilson, W.J., Barrett, A.H.: 1972, Astron. Astrophys. 17, 385.
Wood, P.R., Cahn, J.H.: 1977, Astrophys. J. 211, 499.

GRAIN MANTLE PHOTOLYSIS: A CONNECTION BETWEEN THE GRAIN SIZE DISTRIBUTION FUNCTION AND THE ABUNDANCE OF COMPLEX INTERSTELLAR MOLECULES

J.M. Greenberg
Laboratory Astrophysics, Huygens Laboratorium, University of Leiden, The Netherlands

ABSTRACT

The energy stored as free radicals in the process of growth and photoprocessing of interstellar grain mantles by ultraviolet radiation in clouds is treated as the primary cause of both grain mantle destruction and gas molecule production in dark clouds. The triggering mechanism for energy release is provided by grains colliding with each other as a result of turbulence in the cores of the clouds. A quantitative derivation of a size distribution function of the general form $e^{-\alpha^3 a^3}$ for the mantles of the core-mantle interstellar dust grains, which best determines simultaneously the mean interstellar extinction and polarization curves, is based on the dust-dust collision rate and a grain evaporation efficiency parameter, f, to be determined by experiments performed in the photochemistry laboratory at Leiden University. The same mantle destruction process which produces the size distribution function is used to provide a numerical estimate of molecule ejection into the gas phase in terms of a parameter, α_M, the effective molecule concentration fraction in the evaporating mantle. The evaporation efficiency factor and the molecule concentration function required in the derivations to give a simultaneously consistent picture of the mean size distribution function and the molecule production rate, specifically for H_2CO, may be inferred to be realistic representations of the laboratory results obtained to date. The molecules ejcted by the explosive process are likely to appear in initially excited states and would be subject to subsequent gas phase interactions.

I. INTRODUCTION

The role of interstellar dust in the formation of molecules as observed in the interstellar gas has been discussed by many authors. The principal studies have been made on the formation of H_2 and of more complicated molecules on grain surfaces. For a recent review

Bengt E. Westerlund (ed.), Stars and Star Systems, 173–193.

see Watson (1976, 1978). It is the purpose of this paper to show that the bulk (not the surface) of the grain materials must undergo photochemical processing in the interstellar medium and that there are mechanisms for both manufacturing and releasing many molecules inte the gas from the solid grain which may be subjected to laboratory investigations. We shall start with a basic grain model and consider the consequences resulting from interactions of the interstellar radiation field with the atoms and molecules accreted to form the grain mantles from interstellar gas clouds. An interconnection between the nature of the particle size distribution and the molecule formation rate will be demonstrated. Finally the laboratory analogue of the photoprocessing will be outlined.

II. CLOUD PARAMETERS

a. Basic Grain Model

The interstellar grains appear basically as a bimodal distribution consisting of "classical" sized grains of $\sim 0.15\ \mu$ in size and much smaller (bare) particles $\sim 0.005\ \mu$ in size. The form of the classical sized grains suggested by the observations (see Greenberg 1978a for a detailed discussion of the model) is a core-mantle structure with the core consisting of a "silicate" type material and an ice-like mantle which has been accreted from the interstellar gas consisting of atoms and molecules made up of the abundant condensible species O, C and N. It is these ices in which the interstellar solid state photochemistry takes place.

We use a cylindrical grain model with size distribution (see the discussion in section IV) for mantles a_m on a single core size a_c

$$n(a_m) = \exp\{-5(\frac{a_m - a_c}{a_i})^3\} \tag{2.1}$$

where a_i is a size (cut-off) parameter on the mantle thickness, $a_m - a_c$, distribution. With indices of refraction characteristic of silicates and ices for the core and mantle, respectively, one arrives at a best fit for extinction and polarization by the choice of parameters for the classical and bare particles

$$\left.\begin{array}{l} a_i = 0.2\ \mu \\ a_c = 0.05\ \mu \end{array}\right\} \rightarrow \bar{a}_m = 0.12\ \mu \tag{2.2}$$

$$a_b = 0.005\ \mu$$

The classical particles produce an extinction law with total to selective extinction

$$A(V)/E(B-V) = 3.2$$

The mean space density of the particles is (Greenberg and Hong 1974, Hong 1975)

$$n_{c-m} = 9.5\ e^{-1} \times 10^{-13}\ n_H$$

$$n_b = 4.6 \times 10^3\ e\ n_{c-m}$$

where e is the elongation of the classical particles and n_H is the number density of hydrogen in all forms. For much of the following it is acceptable for simplicity to represent the grains as spherical core-mantle particles($e = 1$).

The mean temperature of the core mantle grains in gas clouds is in the range 10 K to 20 K. We shall use a nominal value of 10 K in the subsequent consideration unless otherwise stated.

b. Gas

The HI gas density varies from $n_H = 0.2\ cm^{-3}$ (intercloud) to $\sim 10^2\ cm^{-3}$ (standard cloud) to $\gtrsim 10^3\ cm^{-3}$ (dense cool cloud). The temperatures are, respectively, of the order of several thousand degrees, 50 - 100 K and < 50 K. In the first two the hydrogen is mostly atomic; in the latter it is mostly molecular. Whenever we use the term n_H it will mean the number density of hydrogen "atoms" in whatever form. We make the usual basic assumption that the overall abundances of the elements in the gas and dust together are similar to that of the solar system (so-called cosmic abundances).

Using the above model for the dust and applying it to the observations of ζ Oph we find (Greenberg 1974, 1978a) that the total number of accountable atoms of O, C and N along the line of sight as either atoms or ions (Morton 1974) or in dust is only about 50 % of those predicted by cosmic abundances. Since the most abundant molecule, CO, supplies at most the order of an additional 20% of C or 10% of O (and nothing to N), there is implied a substantial gas fraction in undetected molecules. This sea of condensable material is actually required to produce the observed growth of grains as indicated by modified optical properties in dense clouds relative to the mean (Carrasco *et al.* 1973).

The average cloud density of gas relative to dust opacity appears to be given by (Morton 1974)

$$N_H/E(B-V) = 4.2 \times 10^{21}\ \text{atoms cm}^{-2}\ \text{mag}^{-1}.$$

This correlation is presumed to hold *almost* universally.

If we combine this with the average ratio of extinction to color excess A(V)/E (B-V) = 3.2 we obtain

$$A(V) = \Delta m_V = 0.71\ N_H \times 10^{-21}\ cm^2 \tag{2.3}$$

which, with an "average" hydrogen density of $n_H = 1\ cm^{-3}$ implies a characteristic extinction in the plane of the Milky Way of about

$$\Delta m_V/D = 2\ mag/kpc$$

It should be noted that E(B-V) does not provide a unique value of the *number* of dust grains because it is dependent on their size which varies from region to region.

c. Radiation Field

We shall be concerned with the effects of ultraviolet radiation on the grains in HI regions. Although we use specifically the Habing (1968) radiation field only modest changes in our results may be expected as compared with other choices we might have made. It is convenient to approximate the energy density in the Habing field by

$$u = 40 \times 10^{-18}\ erg\ cm^{-3}\ Å^{-1} \qquad 2000\ Å \geq \lambda \geq 912\ Å \tag{2.4}$$

which implies a number density of photons between E and E + dE of

$$n(E)dE = 0.029\ E^{-3}dE \qquad 6 \leq E \leq 13.6\ eV$$

where E is in electron volts (eV).

The total number density of photons with energies in this range is then

$$n_{u.v.} = \int_6^{13.6} n(E)dE = 0.003\ cm^{-3}$$

III. GROWTH AND PHOTOLYSIS

a. Collision Rates

Two basic collision rates on the grains are by the gas atoms and molecules and by the ultraviolet photons. Given an atom (or molecule) with number density n_A (where A is the molecular weight) at kinetic temperature T, the molecular or atomic collision rate with a spherical grain of radius a_m is

$$\frac{dN_A}{dt} = n_A\ v_A\ \pi\ a_m^2 \tag{3.1}$$

If n_A includes all the condensible species in the gas (ΣA) and if every such atom or molecule sticks to the grain (this will be discussed further in section IIIc and V) the grain mantle growth rate may be estimated as (Greenberg 1978a)

$$\frac{da_m}{dt} = 3.43 \times 10^{-21} \; n_H \, T^{1/2} \text{ cm s}^{-1} \quad (3.2)$$

where we have used a mean molecular weight $\bar{A} = 16.5$ and a sticking coefficient of unity. Note, that implicit in this derivation is the assumption that the sticking of a molecule like CO is equivalent to the separate sticking of an O and a C atom.

When an atom or molecule sticks it is first a surface molecule but eventually becomes integrated into the volume of the grain. This integration time, i.e. the time required to add one layer $\Delta a_m \simeq 2 \times 10^{-8}$ cm, is then $2 \times 10^{-8}/(da_m/dt) \approx 4 \times 10^3$ yrs for $n_H = 50 \text{ cm}^{-3}$, $T = 50$ K.

The very small bare particles are presumed not to accrete mantles for a number of physical as well as observational reasons.

The ultraviolet photons which are likely to break a chemical bond in the grain mantle are conservatively estimated to have energies greater than 6 eV ($\lambda \lesssim 2000$ Å). Typical bond dissociation thresholds (Calvert and Pitts 1966) for some molecules (with the bond specified in brackets) are in eV: CO {C ≡ O} 11.4; H_2CO {C = O} 7.59; H_2CO {H-C} 3.82; CH_4 {C-H} 4.51; H_2O {O-H} 5.16; NH_3 {N-H} 4.47; CO_2 {O = C} 5.51. In unshielded HI regions the collision rate of such photons is

$$\frac{dN_{u.v}}{dt} = n_{u.v} \, c \, \pi \, a_m^2$$

$$= (3 \times 10^{-3}) \, (3 \times 10^{10}) \, \pi \, a_m^2 \quad (3.3)$$

$$= 5 \times 10^{-2} \text{ s}^{-1}$$

on a 0.12 μm size grain, where c = velocity of light.

The ratio of the photon collision rate to the gas collision rate is

$$\left(\frac{dN_{u.v}}{dt} \Big/ \frac{dN_{\Sigma A}}{dt}\right) = \frac{n_{u.v} \, c}{n_{\Sigma A} \, v_A} \quad (3.4)$$

In a cloud of density n_H we may estimate $n_{\Sigma A} \simeq 10^{-3} \, n_H$ from cosmic abundance arguments. At a gas temperature of T the mean molecule velocity is

$$v_A \simeq \left(\frac{T}{100}\right)^{1/2} \left(\frac{m_H}{m_A}\right)^{1/2} 1.5 \text{ km s}^{-1} \simeq 0.21 \text{ km s}^{-1}$$

for a mean molecular weight of 25 and a temperature T = 50 K. For these values the ratio in Equation (3.4) is $3\, n_H^{-1}\, c/v_A = 4.5\; 10^6\, n_H^{-1}$. It may be shown that equal rates are achieved as a result of the ultraviolet attenuation in the interior of a dark (dense) cloud of the order of 1 pc radius and density $n_H = 10^3\ cm^{-3}$ (Greenberg 1976).

In a cloud with density $n_H \gtrsim 10^3\ cm^{-3}$ it seems certain that just about all molecules stick which strike the grain surface (Salpeter and Watson 1973). In this case one may show that all the gas will have accreted on the grains in a time of the order of

$$\tau_{ac} = \frac{A^{1/2}\, N_H/A(V)}{v_H\, n_H} = \frac{2.6 \times 10^9}{n_H}\ \text{yrs} \qquad (3.5)$$

Another important time scale is the free-fall, or cloud contraction time as given by

$$\tau_{ff} = \left(\frac{3}{4\pi G\rho}\right)^{1/2} = \frac{4 \times 10^7}{n_H^{1/2}}\ \text{yrs} \qquad (3.6)$$

The potential importance of ultraviolet photoprocessing of the grains is immediately demonstrated from the shortness of the time it takes for the photon collision number to equal the total number of molecules in a grain. We may estimate this time as

$$\begin{aligned}\tau_{u.v.} &= N_M/(dN_{u.v}/dt) \\ &= \frac{4\pi}{3}(a_m/d_M)^3/(dN_{u.v}/dt) \\ &\simeq 200\ \text{yrs}\end{aligned} \qquad (3.7)$$

where N_M = number of molecules in a grain, d_M is a mean molecular diameter $\simeq 3 \times 10^{-8}$ cm. It is clear that the photodissociation efficiencies would have to be extremely small for the photoprocessing time scale to approach the cycling time of a grain mantle ($\simeq 10^7$ yrs). The lifetime of the entire core-mantle grain is probably $\simeq 5 \times 10^9$ yrs as derived from an estimated star formation rate (Greenberg 1978b).

b. Ultraviolet Photon Penetration

When a photon strikes the grain it may interact with the surface or be absorbed in the interior. An important question to be answered is the relative proportion of these two phenomena. It seems reasonable to estimate this from the wave-particle duality of the electromagnetic scattering process which gives the probability density for photon interacting as proportional to the squared electric field, $|E|^2$, at all points in the grain. Such calculations have been made (Tielens, unpublished) using Mie theory for homogeneous spheres and for core-mantle spheres. An example relevant to the ultraviolet absorption is shown in Figure 1 where we have chosen to represent qualitatively

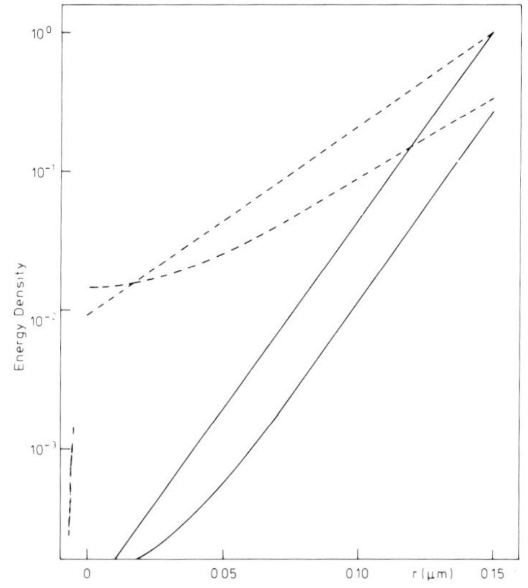

Fig. 1. Energy density distribution in homogeneous absorbing spherical grains of radius 0.15 μm and index of refraction 1.33 - 0.5 is subjected to a bath of ultraviolet radiation. Upper dotted curve and upper dashed curve are exact Mie theory results for λ = 1000 Å and 2000 Å respectively. Straight lines are the pure exponential decay $\exp\{-4\pi m'' t/\lambda\}$ where t is the depth of penetration.

the core and mantle by complex indices of refraction typical of silicates and ices beyond $\lambda \simeq 150$ nm. The calculation shows that the photon penetration follows very closely the exponential decay given by the one-dimensional result $|E|^2 \sim |E_o|^2 \exp(-4\pi m'' t/\lambda)$. The field inside is indeed reduced relative to the surface but the important result one can derive is that the ratio of the number of photons absorbed by surface molecules to those absorbed by interior molecules is only about 10^{-2}. This result is based on the definition that the surface consists of one or two molecular layers and may be qualitatively estimated from $(3\ \Delta a/a_m)e^{4\pi m'' t/\lambda})$ where Δa is the surface thickness, t = mean penetration depth. For Δa = 2 Å, $a_m = 0.15\ \mu$, $m'' = 0.5$, $\lambda = 0.1\mu$,

$t = \frac{a_m}{4} < a_m$, we find the surface to volume photon absorption ratio is $\simeq 0.03$. This turns out to be an overestimate. The Mie theory results obtained for the same m'', a_m, and Δa are 0.01 for either a homogeneous sphere or a core-mantle sphere. We conclude from this that for most classical grains the principal *photolysis* takes place *within the mantle* rather than on the surface of the mantle.

The above result is relevant to the problem of photodesorption of physically bound molecules which have not yet become chemically bonded to the grain. The typical photodesorption energies are considerable less than the bond breaking energies. The grain absorption at these longer wavelengths is therefore considerably smaller so that a far larger proportion of the photons penetrates the grain mantle. This greatly reduces the photodesorption yield. The experimental yield results of L.T. Greenberg (1973) show just such an effect. He obtains yields $\{Y = (\frac{\text{molecules out}}{\text{photons in}})\}$ generally less than 10^{-6}. If we multiply this yield factor times the ratio of photon collision rates with gas collision rates shown in section IIa we find that,

at $n_H \gtrsim 10$, $\frac{dN_{u.v.}}{dt} / \frac{dN_A}{dt} < 1$. In the cases of interest to us here

it therefore seems reasonable to neglect the effects of photodesorption. This will be further discussed in the laboratory analogue section.

c. Photoprocessing of Grain Mantles

We have already shown that there exists a sufficient flux of U.V. photons in the general interstellar medium to provide the basis for considering effects in the interstellar grains. The sequence of events which we call photoprocessing consists of:
(1) formation of trapped free radicals in the low temperature grain by penetrating ultraviolet radiation, (2) gradual build-up of radical concentration during which some of the radicals formed may combine either with each other or with saturated molecules in the mantle, (3) generation of chain reactions of radicals after the build-up to a critical number density, (4) heat energy generated by the chain reaction adequate to evaporate or explode the grain mantle. We summarize here some simple theoretical results of a theory which has been given earlier by Greenberg (1976, and references therein) in order to apply them to a model for mantle destruction and molecule formation.

It turns out that at about 1% stored radical density, the energy released by all the reactions is of the order of that required to evaporate the mantle. Since free radicals recombine with zero activation energy, they can exist no closer than as next nearest neighbors in a matrix. At such a concentration they would constitute about 10% of the total. However such concentrations are statistically unrealistic because local heating generated by recombinations will eliminate this mathematical possibility and it is not likely that stable concentrations greater than $\sim 1 - 2\%$ may occur often. The net efficiency for free radical production - taking into account that some radicals are recombining during irradiation - is certainly less than unity. If we estimate it conservatively as $\varepsilon \gtrsim 10^{-3}$, then the time required to achieve the desired 1% of radical density is

$$\tau_{1\%} = 10^{-2}\ \varepsilon^{-1}\ \tau \simeq 1.7 \times 10^3\ \text{yr}$$

which is still quite small compared with the grain cycletime of 10^7 years. As a matter of fact we envisage that during its lifetime a grain will have undergone a large number of build-ups and partial recombinations (not leading to enough energy for evaporation) so that the repeated break up and recombination processes may lead to the production of quite complex molecules and possibly polymerization of the mantle.

The probability for spontaneous generation of a chain reaction and the release of the total radical stored energy is greatly enhanced by an impulsive temperature rise in a grain produced by, say, the collision of two grains. We are not picturing here the high velocity collisions required to evaporate normal materials. As a matter of fact, from some preliminary laboratory studies (see section IV) there is an indication that the temperature increase required is only to $T \simeq 50$ K. Such a temperature is achieved by grain-grain collisions at only about 0.1 km s^{-1} - the precise value depending on the low temperature specific heat of the mantle. This velocity is much less than the 10 to 20 km s^{-1} which had first been discussed by Oort and van de Hulst (1946) and reviewed recently by Salpeter (1978).

We thus picture a typical grain mantle as spending a major part of its lifetime consisting of a mixture of complex and simple molecules with embedded frozen radicals for which the stored energy may occasionally be released so completely as to totally evaporate the mantle material. More frequently there may be reactions which produce local hot spots in a mantle some of which eject some surface molecules, but many will merely lead to local modifying of the molecular composition. It is interesting to note that any of the stored radicals, located at or near the surface of the grain provide the possibility for chemical interaction with accreting atoms and molecules which should be included in a consideration of the growth as well as the chemical composition of grains. The grain may look, to an incoming molecule, like a partially shaved porcupine with the needles being unpaired electron clouds sticking out. These active sites would be only about 2 atoms apart so that migration and chemical combination by a colliding molecule should be quite rapid.

A schematic illustration of the chemical evolution of a grain during one cycle of its history is shown in Figure 2.

IV. MANTLE EXPLOSIONS, MOLECULE PRODUCTION AND THE GRAIN SIZE DISTRIBUTION

In this section we propose to demonstrate a high degree of internal consistency between such apparently diverse questions as the grain size distribution function and the rate of formation of complex molecules in dense clouds. The one basic assumption will be that in dense

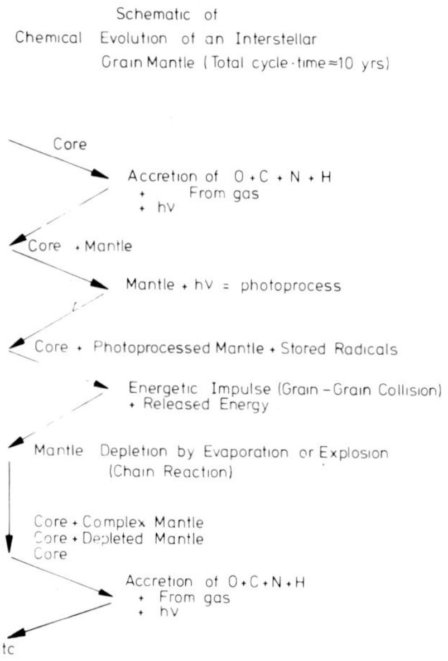

Fig. 2. Schematic of chemical evolution of a grain mantle.

molecular clouds there exists a degree of internal turbulence (sub clouds) adequate to provide a significant number of collisions between grains at velocities of the order of 0.1 km s^{-1}. The widths of molecular lines have even been used to infer much higher internal velocities (Kwan 1978; Leung 1978).

a. The Size Distribution Function

It has been shown (Hong and Greenberg 1978) on empirical grounds that among a variety of one parameter size distributions, a distribution of interstellar grains with general characteristics similar to exp $\{-\alpha^3 a^3\}$ (see section IIa) best yields the most commonly observed combination of the total to selective extinction ratio $R = A(V)/E(B-V) \simeq 3$, and the maximum polarization wavelength, $\lambda_{p_{max}} \simeq 5500$ Å.

Since the exp $\{-\alpha^3 a^3\}$ is clearly favored it is useful to recall its physical justification. The differential equation for grain growth and destruction is

$$\dot{a}\,\frac{dn(a)}{da} + D(a)n(a) = 0 \qquad (4.1)$$

where n(a)da is the number of particles between radius a and a+da, $\dot{a}$ is the rate of growth (assumed constant) by accretion of mostly C, N and O atoms and molecules from the interstellar medium, and D(a) is the probability per unit time that a particle of size a is destroyed. With the assumption that D(a) is proportional to the grain area

$$D(a) = Ka^2, \tag{4.2}$$

we obtain the solution (Greenberg 1966)

$$n(a) = n_o \exp(-\alpha^3 a^3), \tag{4.3}$$

$$\alpha^3 = K/3\dot{a}$$

The value of α^3 implied by our standard core-mantle model (a_i = 0.2μm, Eq. (2.2)) is $\alpha^3 = 5/(0.2)^3 = 625\ \mu m^{-3}$. Combining this with the value of $\dot{a}$ averaged over several grain cycling times (so that n_H is replaced by $\langle n_H \rangle$), and with a mean temperature T = 100 K, one obtains $K = 3\dot{a}\alpha^3 = 2 \times 10^{-7}\ \langle n_H \rangle$. The destruction probability per unit time for the mantle of a grain with radius $\bar{a}_m$ = 0.12 μm is then $D = Ka^2 = 0.03\ \langle n_H \rangle$ per 10^7 yrs. Thus the mean lifetime of a grain (subject to destruction $\sim a^2$) required to satisfy the best size distribution is given by

$$\tau_D^{-1} = 0.3 \times 10^{-8}\ \langle n_H \rangle\ yr^{-1} \tag{4.5}$$

If we use a mean hydrogen density of 1 cm^{-3} (assuming the relation that the time average for growth equals the space average and that growth takes place at any density) we find that about one grain in ten loses its mantle each $10^{7.5}$ years.

This concept of mean density requires some discussion. A fraction (see section IVd) of the interstellar gas is concentrated into clouds whose density is so high and the obscuration so large that not enough is known about those dust grains which are larger in size than average (see, for example, Carrasco *et al.* 1973). Therefore, our mean grain size is obtained over an incomplete sample of space and may be an underestimate. If so, then the value of a_i in Equation (2.2) is too low. This means that α should be reduced by the factor $(a_i/a_i')^3$ (where a_i' includes those dust grains which are in dense clouds) if we use $\langle n_H \rangle \simeq 1$ which includes *all* the gas. The destruction probability, however, is $\sim \bar{a}_m^2$ which should be replaced by $\bar{a}_m'^2 > \bar{a}_m^2$. The ratio $\bar{a}_m'/\bar{a}_m$ is less than a_i'/a_i because it includes a constant core. Therefore $(\bar{a}_m'/\bar{a}_m)^2\ \alpha'^3 \lesssim (a_i'/a_i)$ - but still > 1.

On the other hand the value of T = 100 K used in our calculation of a is characteristic of the teneous clouds and should be replaced by the lower value $T' \simeq 50$ K, characteristic of the dense clouds. In conclusion, it would appear that τ_D in Eq. (4.5) may be an upper bound to the destruction rates.

b. Dust-Dust Collisions in a Molecular Cloud

Let us use v_d as the effective velocity for dust collisions leading to explosions. The time τ_{dd} is defined as the time it takes for a dust grain to traverse a path for which the total optical depth (proportional to grain area) is equal to unity. We shall ignore, in this calculation, all factors of order unity. We obtain

$$\tau_{dd} = \frac{0.67 \times 10^{14}}{n_H v_d} \text{ yr} \tag{4.6}$$

where we have used the standard gas to dust ratio given in section II.

The mean fraction of dust-dust collisions which effectively totally evaporates a grain mantle at v_d will be called f. The actual averaging must be quite complex because it involves not only an average over a velocity spread as well as over impact parameters, but also includes some partial evaporations. Dust-dust collisions resulting from cloud-cloud collisions are not included in our considerations because they appear to produce a smaller *mean* rate than given by Eq. (4.6) (see, e.g. Salpeter 1977). However, effects at the cloud collision edges may be at least locally significant in molecule production and fluorescence (section IVc).

c. Dust-Dust Collisions and the Production of Molecules

A problem of some concern in dark clouds is to understand the presence of large numbers of molecules where the accretion rate of these molecules on the grains might be expected to deplete them rapidly (Watson 1976, Greenberg 1978c). If the grains themselves are a significant source of such molecules then, because both the accretion rate and production rate are proportional to the grain density, this question might be easily answered.

The rate of production of a particular molecule M resulting from dust-dust collisions in a dark cloud is given by

$$\frac{dn_M}{dt} = \frac{dn_d}{dt}\left(\frac{n_M}{n_d}\right) \tag{4.7}$$

where the ratio n_M/n_d is the effective number of the molecules M released from each grain mantle. This number is of the order of

$$(n_H/n_d) = 4/3\ \pi\ (a/d)^3\ \alpha_M \tag{4.8}$$

where d is the size of the molecule and α_M is a factor (generally << 1) which includes not only the instantaneous number of released molecules M but also those which may be subsequently generated by (sufficiently rapid) dissociative reactions in the gas.

Thus the net rate of molecule production is

$$\left(\frac{dn_M}{dt}\right)_{dd} = \left(\frac{n_d}{\tau_{dd}} f\right) \frac{4}{3} \pi \ (a/d)^3 \alpha_M \tag{4.9}$$

$$= 1.85 \ n_H^2 \ 10^{-10} \ f\alpha_M$$

where we have used $n_d = 10^{-12} \ n_H$ (see section IIa) and $v_d = 10^5$ cm s^{-1}.

We now equate the production rate for the molecule to the loss (accretion) rate - assuming that accretion is the dominant cause of depletion in a dark cloud rather than, say, photodissociation (Greenberg, 1973) - in order to arrive at a steady state, given by

accretion = formation

$n_M/\tau_{ac} = (db_M/dt)_{dd}$

which becomes, using Equations (2.5) and (4.7),

$$n_M \ n_H = 4.8 \ n_H^2 \ 10^{-2} \ (f\alpha_M) \tag{4.10}$$

Applying this to H_2CO in a dark cloud, we let $n_M = n_{H_2CO} \approx 0.4 - 0.8 \times 10^{-8} \ n_H$ with the result that

$$f\alpha_{H_2CO} \simeq 10^{-8} \tag{4.11}$$

This is a satisfying result in that it does not require an exceptionally large fraction of the ejected mantle molecules to be specifically H_2CO (letting $f \simeq 10^{-2}$ as estimated in the following). It is probable that many molecules explosively ejected from grain mantles are in highly excited states even though the cloud temperature is low.

d. Dust-Dust Collisions and the Mantle Size Distribution Function

If the major loss of grain mantles results from dust collisions in dense clouds then, in some way, this must be related to the size distribution factor D. The first thing we realize is that the rate of mantle destruction in the dark cloud is not the same as the *mean* rate of mantle destruction given by D because it does not act in a continuous fashion but rather occurs (or is effective) in only a portion of each grain cycle.

The assumption of the cloud turbulent velocities is used only in the dark cloud phase, and it is therefore only from the time this phase begins that we need consider the mantle destruction. The total lifetime of the dark cloud starting from this point is at least as great as the free-fall time so that we may estimate a lower bound to the fraction of the grains colliding by taking the ratio of the collision rate to the free-fall rate. We see that this gives

$$\tau_{dd}^{-1}\ \tau_{ff} = 0.6 \times 10^{-1}\ n_H^{1/2} \tag{4.12}$$

which implies that, beyond the density $n_H = 10^4\ cm^{-3}$, the probability is of order of unity that subsequently all grains will have collided at least once.

Although the interstellar gas and dust in HI regions is distributed over a wide range of densities, it is strongly concentrated in the CO clouds where the density is $n_H \gtrsim 10^3$. It has been variously estimated that the fraction of all the interstellar gas in such clouds is at least 1/2 (Burton 1977) and perhaps as much as 9/10. We shall use a fraction 3/4 as an intermediate of these two. We again assume that the space average is a good measure of the time average. Therefore we may estimate the mean dust-dust collision (rate) over a grain cycle time by first calculating the collision rate at those densities for which turbulence is effective and then multiply this rate by the fraction of the total mass concentrated at such densities. It is difficult to be very precise about either the density at which turbulence sets in or about the fraction of the CO cloud mass which exists at this density. However, since the final result is proportional to a product involving these two choices there is a degree of self cancelling of the errors we make in the individual terms.

Suppose we denote the fraction of CO clouds with densities $n_H \geq n_c$ at which turbulence sets in as β_c, and the fraction of dust-dust collisions which lead to grain mantle destruction at and beyond this density as f. Then the mean rate of mantle destruction over a cycle time, which we call D_{dd}, is

$$D_{dd} = \beta_c\ f(0.75)\ \tau_{d-d}^{-1}\ n_c \tag{4.13}$$

Equating this to the empirical destruction factor D we get

$$\frac{(0.75)\ \beta_c f\ n_c}{0.67 \times 10^{10}} = 0.3 \times 10^{-8}\ yr^{-1}$$

from which the efficiency factor f is

$$f = 0.7\,(\beta_c\, n_c)^{-1}\, 10 \qquad (4.14)$$

For $n_c = 10^3\ cm^{-3}$ (which means $\beta_c = 1$) one gets $f \simeq 0.01$.

It may well be that n_c is more like $10^4\ cm^{-3}$. Since we do not know what fraction of the CO clouds are at density 10^4 we can only state that if the factor $\beta_c\, n_c$ is roughly constant, the value of f is still $\sim 10^{-2}$. However, the possibility that $f \simeq 0.1$ is not excluded. In either case we find the not obviously unrealistic criterion that only one dust-dust collision in 100 (or 10) leads to total mantle evaporation. This statement should be interpreted as including, in the averaging, those more numerous collisions which lead to partial mantle evaporation.

The total rate of energy release in a cloud by dust-dust collisions during the initial stages of turbulence has been estimated to be comparable with the lower limit of cosmic ray heating (Spitzer 1968). No matter in what forms this energy appears it is undetectable.

V. LABORATORY ANALOG

In the Laboratory Astrophysics group at Leiden University we are currently studying the photochemistry of analog interstellar materials at temperatures down to about 10 K. The scientific staff are Dr. L. Allamandola, Dr. F. Baas, Drs. W. Hagen and Mr. C. van de Bult. The basic equipment consists of: (1) a closed system helium cryostat with a cold finger at 10 K on which may be deposited thin (∿ 0.2 μm) samples of various gas mixtures of simple molecules containing oxygen, carbon and nitrogen; (2) a Fourier Transform Infrared spectrometer (Digilab FTS 15) to study the absorption spectrum of the sample; (3) a quadrupole mass spectrometer (mass range 1 - 300 a.m.u.) to study the molecular weights of the molecules evaporated off the cold finger; (4) a set of ultraviolet lamps and windows as sources of the production of the photolyzing photons; (5) a gas handling system for mixing and depositing various molecular mixtures; (6) a visual and vacuum ultraviolet spectrometer and phototube to measure the light output of the sample when it is warmed; (7) various temperature and pressure measuring and control devices. Additional equipment is to be used with the above to study the infrared fluorescence which can be emitted by excited molecules in the mantle (Allamandola and Norman 1978). We will be studying the visual and ultraviolet absorption spectra of the irradiated solids as well as the infrared absorption and emission. Some of them will be reported on by Dr. Allamandola *et al.* A detailed description of some of the laboratory and its operation will be reported on at this meeting by Dr. Hagen *et al.* All I will do here is to describe as illustrative the results of one of the first experiments which was performed in our laboratory. A deposit of NH_3 and CO was condensed on the cold finger in the cryostat. The infrared spectrum of this layer was then recorded. The ultraviolet photons from a hydrogen lamp with a Mg F window were then used to photolyze the sample for 2 hours. The infrared spectrum was again recorded. Positive identification of the newly created radicals and molecules are listed in Table 1. There are probably more complicated radicals and molecules but in insufficient number to have been readily detected.

Table 1. Identified molecules and radicals in photolyzed sample of NH_3 + CO (1 ÷ 100) plus trace H_2O, CO_2 at 10 K.

Molecule/Radical	Frequency
HCO (Formyl)	2488, 1860, 1090
NH_2	1499, (1506), (3213)
H_2CO (Formaldehyde)	2865, 2796, 1737, 1506, (1499)
HOCO	(1830)
CO_2	2348, 660
HNCO	2263, 815, 810, 592
NCO	1936
$HCONH_2$ (Formanide)	3529, 1726, (1265)
Unassigned features:	2040, 1890, 1876, 1855, 1794, 1719, 1506, 1199, 1105 (Milligan and Jacox, J. Chem. Phys. 1965, 43, 4487; 1971, 54, 927.)

When the irradiated sample was allowed to warm up there appeared a greenish-blue fluorescence which continued for some time at about 25 K. This fluorescence subsequently died down and upon continued heating another fluorescence appeared at about 30 - 35 K (Fig. 5). The sample was then heated slowly to about 45 K at which time a flash of light appeared. This experiment was performed four times with varying temperature histories of the post-irradiated samples. The samples were either cooled back to 10 K between fluorescence or allowed to warm up continuously to the 45 K. In each case the flash of light was accompanied by a sharp pressure rise. We believe this to be presumptive evidence of the kind of explosive process which can lead to large molecule formation and ejection into the interstellar medium by interstellar grains which are participating in the star formation process and which have been heated by the kinetic gas temperature or by grain-grain collision in the contracting cloud or by a protostellar source of energy. A very preliminary estimate of the optical energy released during fluorescence shows it to be consistent with the idea of ∿ 1% stored radicals in the system.

A series of infrared absorption spectra of a laboratory sample taken before irradiation and after irradiation is shown in Figure 3 where the formation of new radicals and molecules is clearly demonstrated. Fig. 4 shows how, on a relative scale, the changes in density of molecules and radicals occur during ultraviolet radiation and following warm up.

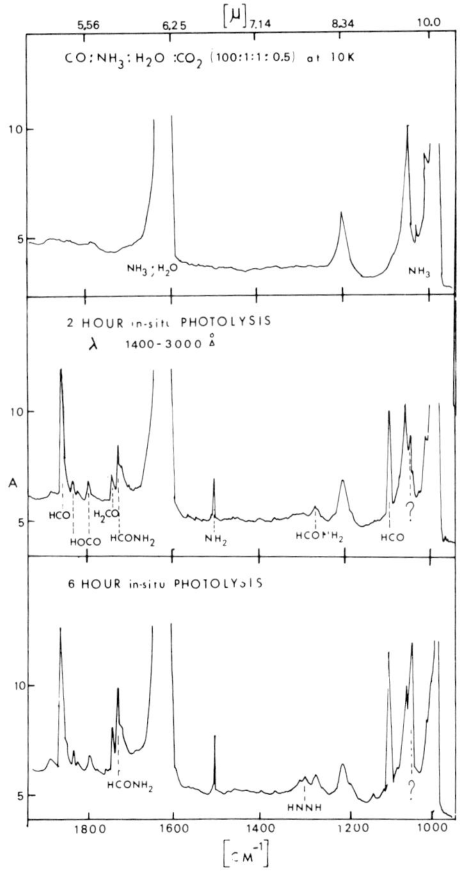

Fig. 3. Infrared absorption spectra of a molecular mixture before ultraviolet irradiation with a hydrogen lamp (upper) after two hours irradiation (middle) after 6 hours irradiation (bottom).

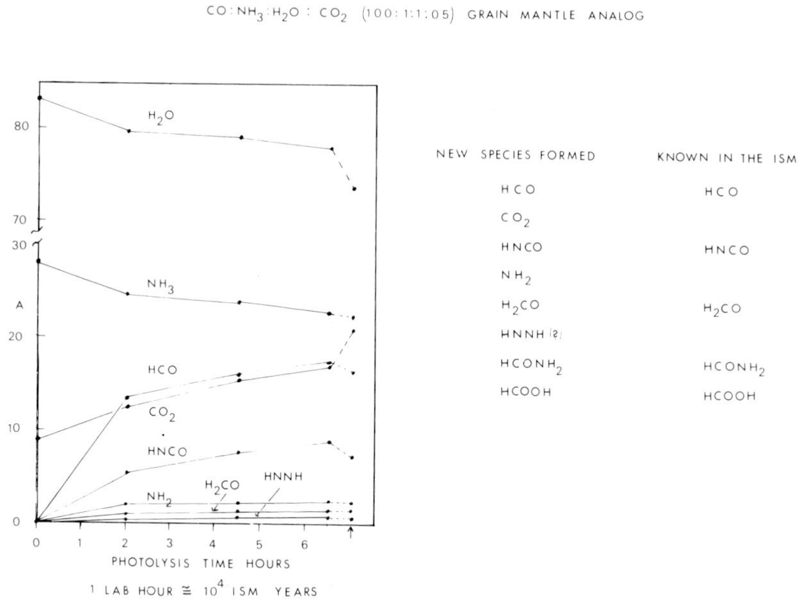

Fig. 4. Change of chemical composition produced by photolysis (see Figure 3). Arbitrary vertical scale.

Should either the stored energy of the grains be too small or the chain reaction not go to completion the recombined radicals and molecules which remain within the grain will be of the larger or polymerized variety and consequently the residual mantle on the dust will tend to be substantially more refractory than the classical ices consisting of the relatively volatile substances H_2O, CH_4 and NH_3.

VI. CONCLUDING REMARKS

It has been shown that dust-dust collisions in turbulent dense clouds can provide the triggering mechanism for releasing the stored energy in previously photoprocessed grains. The rate of ejection of molecules from the resulting explosion or evaporation of grain mantles appears adequate to account for the appearance of such complex molecules as H_2CO as well as larger ones. The mantle ejection process compensates for the loss of molecules by sticking on grains.

It is expected that molecules produced by an explosive process

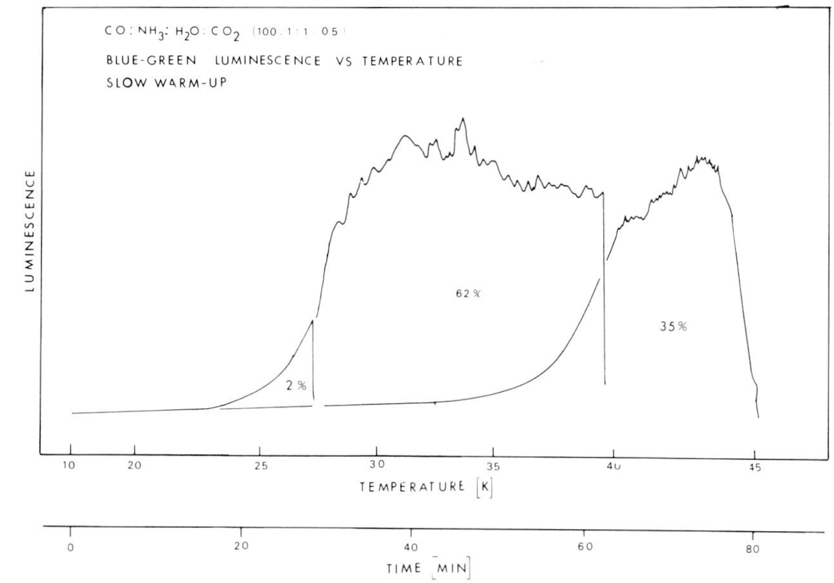

Fig. 5. Luminescence during warm-up of photolyzed material from 10 K to 45 K.

will be in highly excited states so that further gas phase reactions should be included in the final predictions of concentrations as a function of cloud density which may be compared with observations.

Additional investigation should also be made on the effects produced by the very small bare particles in the bimodal size distribution.

The work presented here has been based on a collaboration within the Laboratory Astrophysics Workgroup at the University of Leiden. More detailed accounts of some of the material in this paper will appear elsewhere.

VII. ACKNOWLEDGEMENTS

I should like to acknowledge the important contribution by Dr. L. Allamandola, Dr. F. Baas, Mr. C. van de Bult, Drs. W. Hagen, Dr. C. Norman and Drs. A. Tielens in developing both the theoretical and experimental results. I am also indebted to Dr. S. Hong for some important ideas.

REFERENCES

Allamandola, J.L. and Norman, C.A.: 1978, Astron. Astrophys. 63, L23.
Burton, B.: 1977, Discussion following review by F.J. Kerr, *Star Formation,* IAU Symp. No. 75, p. 35, Ed. T. De Jong, A. Maeder, D. Reidel.
Calvert, J.G. and Pitts, J.N., Jr.: 1966, *Photochemistry,* John Wiley and Sons.
Cameron, A.G.W.: 1973, Space Science Reviews, 121, D. Reidel Publ. Co., Dordrecht, Holland.
Carrasco, L., Strom, S.E. and Strom, K.M.: 1973, Astrophys. J. 182, 95.
Goldanski, V.I.: 1976, Ann. Rev. Phys. Chem. 27, 85.
Greenberg, J.M.: 1966, Proc. of IAU Symp. No. 24: *Spectral Classification and Multicolor Photometry,* p. 291, Eds. K. Lodén, L.O. Lodén and U. Sinnerstad, Academic Press, London.
Greenberg, J.M., Yencha, A.J., Corbett, J.W. and Frisch, H.L.: 1972, Mém. Soc. R. Sci. de Liège, 6e série, tome III, pp. 425-436.
Greenberg, J.M.: 1973, in *Molecules in the Galactic Environment,* Proc. Symposium on Interstellar Molecules, Oct. 1971, p. 93, Eds. M.A. Gordon and L.E. Snyder, John Wiley and Sons, London.
Greenberg, J.M.: 1974, Astrophys. J. 189, L81.
Greenberg, J.M.: 1976, Astrophys. Space Sci. 39, 9.
Greenberg, J.M.: 1978a, *Cosmic Dust,* Chapt. 4, p. 187, Ed. A.J.M. McDonnell, J. Wiley and Sons, London.
Greenberg, J.M.: 1978b, Review presented at IAU Colloquium on *Protostars and Planets,* Tucson, Arizona 1978 (To be published in the Moon & Planets).
Greenberg, J.M.: 1978c, Paper presented at 21st Liège Astrophysical Symposium on *Small Molecules,* in June, 1977.
Greenberg, J.M. and Hong, S.S.: 1976, Astrophys. Space Sci. 39, 31.
Greenberg, L.T.: 1973 *Interstellar Dust and Related Topics,* Proc. IAU Symp. No. 52, p. 413, Eds. J.M. Greenberg and H.C. van de Hulst, D. Reidel Publ. Co., Dordrecht, Holland.
Habing, H.J.: 1968, Bull. Astron. Inst. Ned. 19, 421.
Hong, S.S.: 1975, *Unified Model of Interstellar Grains,* Ph.D. Thesis, State Univ. of New York at Albany.
Hong, S.S. and Greenberg, J.M.: 1978, Astron. Astrophys., in press.
Jackson. J.L.: 1959a, J. Chem. Phys. 31, 154.
Jackson, J.L.: 1959b, J. Chem. Phys. 31, 722.
Kwan, J.: 1978, Astrophys. J. 223, 147.
Leung, C.M.: 1978, Bull. American Astron. Soc. 9, 540.
Milligan, D.E. and Jacox, M.E.: 1965, J. Chem. Phys. 43, 4487.
Milligan, D.E. and Jacox, M.E.: 1971, J. Chem. Phys, 54, 927.
Morton, D.C.: 1974, Astrophys. J. 193, L35.
Oort, J.H. and van de Hulst, H.C.: 1946, Bull Astron. Inst. Ned. 10, 187.
Salpeter, E.E. and Watson, W.D.: 1973, *Interstellar Dust and Related Topics,* Proc. IAU Symp. No. 52, p. 363, Eds J.M. Greenberg and H.C. van de Hulst, D. Reidel Publ. Co., Dordrecht, Holland.
Salpeter, E.E.: 1977, Ann. Rev. Astron. Astrophys. 15, 267.

Spitzer, L.: 1968, *Diffuse Matter in Space*, Interscience Publishers, New York.
Watson, W.D.: 1976, Rev. Month. Phys. 48, 513.
Watson, W.D.: 1978, Review presented at 21st Liège International Astrophysical Symposium, in June 1977.

PROPERTIES OF H II REGIONS

Marcello Felli
Osservatorio di Arcetri, Florence, Italy

1. INTRODUCTION

Symposia and several review papers have recently been devoted to H II Regions (see e.g. the Mittelberg Symposium on H II regions and Related Topics 1975; the I.A.U. Symposium no. 75 on Star Formation 1977; Mezger 1978a, for a discussion of the properties of the interstellar matter; Brown *et al.* 1978, for the interpretation of radio recombination lines emitted by H II regions; Woodward 1978, for the mechanisms capable of creating stars and Panagia 1978, for the IR emission of dust associated with H II regions).

From the above it would appear that the main properties of H II regions are now known fairly well and that their evolution is timed in a more general process which, starting from the collapse of a massive cloud, leads ultimately to the formation of an OB association imbedded in the diffuse remnants of the original H II region. One might then ask what is now the point of studying H II regions, what still remains to be discovered. The answer, as always in these cases, depends on the degree of required accuracy and completeness. For instance, the idealized model adopted until now, in which a star forms at the centre of a molecular cloud and develops a spherically symmetric H II region, which subsequently expands, although it is the simplest and safest approach, finds little correspondence, especially in the later stages, with the extremely complex scenario that faces the observer. Furthermore, in order to obtain general properties one needs a good deal of well-studied and clearly understood cases, particularly considering that we are dealing with regions in different stages of their evolution. Finally, selection effects in the observations should be taken into account, as well as instrumental and theoretical bias which might affect the completeness and uniqueness of the results. To tell the other side of the story, I will not attempt to describe the "general" properties of H II regions, but rather to examine the more recent observational basis for the study of H II regions. In doing so, I shall deal essentially with radio continuum observations with particular regard to the ionization balance in an H II region (sec-

Bengt E. Westerlund (ed.), Stars and Star Systems, 195–220.

tion 2) and to the indications given by H II regions on the phases immediately following star formation (section 3). Clearly this is only one aspect of the properties of these regions and by no means sufficient for a complete discussion of them.

2. THE IONIZATION BALANCE

Recalling a few basic facts, in an equilibrium situation to keep hydrogen ionized in a given region one needs a constant supply of Lyman continuum photons (Ly-c) to balance recombinations. Let N_L^* be the flux of stellar Ly-c photons of the star(s) associated to the H II region, and N_L^R the flux of Ly-c photons absorbed within the H II region. N_L^R can be derived from the continuum radio flux of the region and its distance provided that: a) the radio optical depth τ_{ff} at the frequency of observation is much less than unity (Rubin 1968; Hjellming 1968). With the additional hypothesis b) the exciting star(s) is (are) located inside the H II region and this is completely ionization bounded, c) the region is dust free, all stellar Ly-c photons are absorbed within the H II gas and the two quantities N_L^* and N_L^R are related by the equality:

$$N_L^* = N_L^R \quad . \tag{1}$$

In this way an ionization balance between the stellar supply of Ly-c photons and those absorbed within the H II region can be performed.

When any of the conditions a) to c) are not satisfied it will be $N_L^* > N_L^R$. Alternatively if conditions a) to c) are satisfied but not all the exciting stars in the H II region have been detected it can be $N_L^* < N_L^R$.

Since the relation (1) is one of the basic tools in the study of the relationship between H II regions and exciting stars, I shall now discuss separately the various effects that may limit its use, illustrating them with examples.

2.1 The accuracy of the N_L^* values

Values of N_L^* as a function of the spectral type (Sp) and the luminosity class have been derived by several authors using stellar model atmospheres and recent calibration of absolute magnitudes and temperatures (Rubin 1968; Churchwell and Walmsley 1973; Panagia 1973; Israel, *et al.* 1973; Kazes 1975; Torres-Peimbert *et al.* 1974). The values of Log N_L^* are plotted in figure 1 as a function of S_p. The spread of the N_L^* values is quite noticeable, up to an order of magnitude in some cases and much worse than the possible accuracy of the N_L^R values.

The best way to check which of the theoretical (N_L^* - S_p) relation should be used is, of course, to apply (1) itself to regions where conditions a) to c) are satisfied.

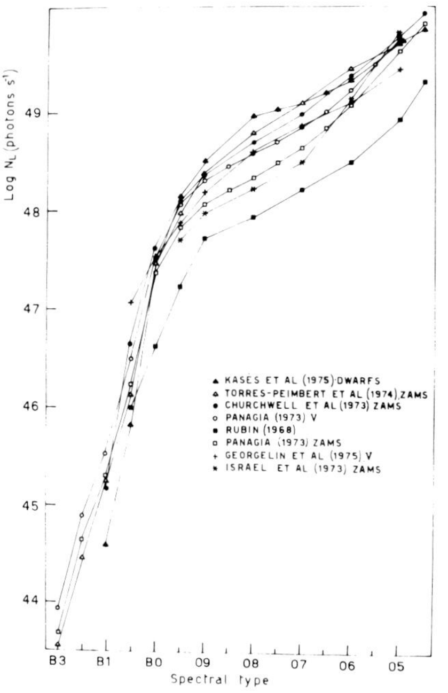

Figure 1: Flux of Ly-c photons (Log N_L^*) as a function of the spectral type (Sp) according to several authors.

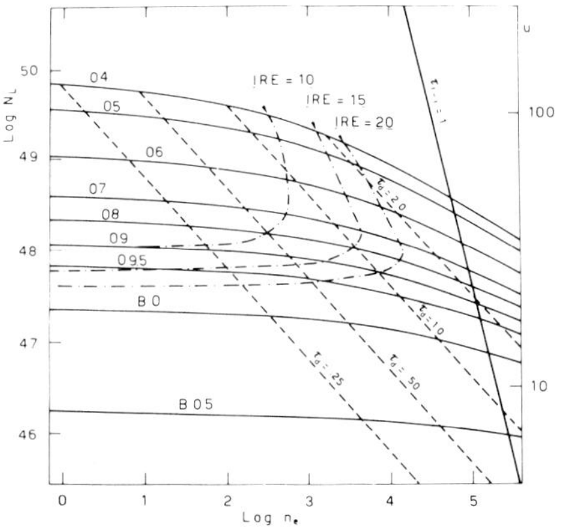

Figure 2: Effects of dust absorption in the Ly-c, from Panagia (1974b). Assumptions: a) uniform distribution of gas and dust; b) $\bar{k}$ (Ly-c) = 6.5 10^{-22} n_H; c) individual Strömgren spheres with ZAMS exciting stars. IRE = $L_{IR}/(N_c' h\nu_\alpha)$. For comparison $\tau_{ff} = 1$ for dust-free nebulae is also shown.

Studies of unselected samples of H II regions had shown a good agreement between N_L^* and N_L^R for later spectral types (08-B0 or cooler star) where the spread of theoretical values is also smaller, whereas the N_L^R were consistently less than the N_L^* for hotter stars. This discrepancy had alternatively been interpreted in terms of density bounded nebulae (Churchwell 1974) or effects of dust absorption in the Ly-c. (Kazes *et al.* 1975).

Georgelin *et al.* (1975, from here on GLM) selected directly a sample of 45 H II regions which fulfilled all the criteria a) to c) and had only one exciting star of known spectral type. To derive an empirical (N_L^* - S_p) relation, they related $N_L^*/\pi F_v$ (the ratio of the ionizing photon flux to the energy emitted in the photometric V band) to the S_p. From this relation, with the additional knowledge of the (M_v-S_p) relation they derived the (N_L^*-S_p) relation shown in figure 1. Given the previous scatter between model computations, the agreement of this relation and Panagia's (luminosity class V) can be considered satisfactory in the range 05-B0, the best being for 07-08 and differences up to 30% in N_L^* for the other spectral types. To check if dust absorption was responsible for the residual difference between the two relations we tried to correlate the dust optical depths τ_d (derived from the comparison of

Panagia's N_L^* with the N_L^R of the regions of the GLM sample) with the column density of ionized hydrogen $n_e R$ (Felli *et al.* 1978a). No correlation was found, because the intrinsic scatter of the values was greater than the studied effects. Nor would the problem be settled by IR observations because, according to the derived τ_d's, the expected IR emission of heated dust would be beneath detection. We conclude, in agreement with Lortet (1975), that dust absorption can be present, but its effects cannot be outlined owing to the very low densities of the regions of the GLM sample and the present accuracy of the data.

2.2. Dust absorption effects

To study the effects of dust absorption in the Ly-c, nebulae which are strong IR sources would be preferred; however in these cases the exciting stars are often heavily obscured and relation (1) cannot be applied in its simple form.

The effects of dust absorption has been dealt with by several authors (Petrosian *et al.* 1972; Panagia 1974a; Balick 1975; Natta and Panagia 1976). To point out its relevance to the present discussion we refer to figure 2 by Panagia (1974b), where the variation of Log N_L^R for ionization bounded spherical nebulae with uniform density is drawn as a function of Log n_e, for spectral types from O4 to BO.5 (ZAMS).

Figure 2 clearly indicates that dust absorption in the Ly-c increases with electron density and is stronger in earlier spectral types. The curves in which τ_d (the dust optical depth in the Ly-c) is equal to 0.2, 0.5, 1.0 and 2.0 are also drawn and, for comparison, the curve in which $\tau_{ff} = 1$. Curves labelled IRE define the zone where IRE = $L_{IR}/N_c^{\prime} h\nu_\alpha$ is equal to 10, 15, 20, and describe the effect of heating of the dust and its emission in the IR with respect to pure Ly-α heating. The reduction of N_L^R due to dust absorption is negligible for low densities but may amount to orders of magnitude at high densities and for earlier spectral types, in which case any use of (1) is pointless.

After the detection of these effects in the compact components in W3 (Wynn-Williams *et al.* 1972), many other examples have been found, some of which will now be briefly discussed.

2.2.1. S140 IR - A selection of different observations of the S140 region is given in figure 3. An unresolved strong IR source (Blair *et al.* 1978) is located in the region of highest T_{CO} of an extended molecular cloud. The molecular cloud is bounded on the SW part by the bright nebulosity produced by the nearby BO star HD 211880. The total luminosity (up to 175 μm) of the IR point source is about 2×10^4 $L_\odot$, compatible to that of a late O or early B star (Harvey *et al.* 1978), and this source is believed to be the main supply of heating to the brighter part of the molecular cloud. Previous upper limits of the radio flux associated to this component were 20 mJy at 5 GHz (Felli *et al.*1978b). Two preliminary radio maps at 1.4 GHz and 0.610 GHz clearly indicate the presence of a small radio source slightly to the south of the IR peak with flux density

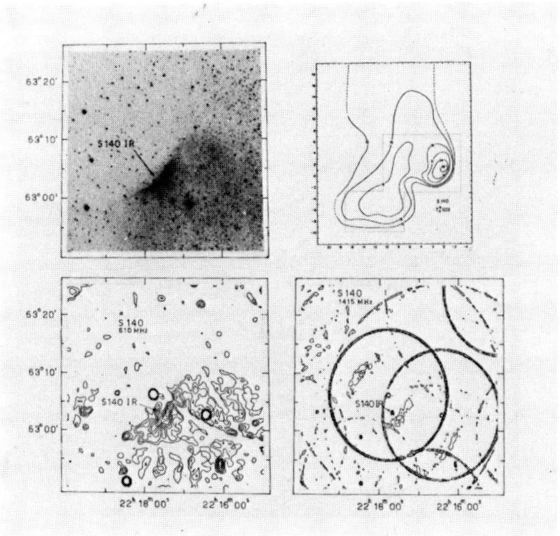

Figure 3: The S140 Region. a) Reproduction of the red Palomar Sky Survey print. b) Contour map of T_A, the corrected antenna temperature of CO. The offsets are in minutes of arc with respect to $\alpha = 22^h17^m42^s$, $\delta = +63^o03'45''$. (Blair et al., 1978; *illustration courtesy of G. Herbig and 'The Astrophysical Journal', published by the University of Chicago Press; © 1977 The American Astronomical Society*). c) Preliminary 610 MHz map of the region. Resolution 60" (Harten 1978). d) Preliminary 1415 MHz map of the region. Resolution 24" (Harten 1978).

compatible to the previous upper limit (Harten 1978). The detection of the radio emission leaves two possible solutions for the S140 IR source: either large amounts of dust mixed with the gas absorb almost completely the Ly-c photons and leave only a small fraction to the gas, or a very dense region ($n_e > 3.10^6$ cm^{-3}) that is optically thick and completely self absorbed. The last solution, however, is not in agreement with the continuum radio spectrum of the source which is apparently flat or even increasing to lower frequencies.

2.2.2. <u>S106</u> - An even more complex situation is to be found in S106, relevant observations of which are grouped in figure 4. Two separated curved structures of ionized hydrogen, pointing to a common origin, can be seen in the radio map of Israel and Felli (1978). They coincide with bright H_α nebulosities (Deharveng 1978) and 12.6 μm structures (Pipher *et al.* 1976). The northern one suffers much more optical obscuration as can be seen from a comparison of radio and optical surface brightness. The entire nebulosity is at the peak of an extended molecular (CO) cloud, slightly elongated in the NE-SW direction, and appears to be in slow rotation around a perpendicular axis, with a $\Delta v \approx 4$ km/sec between the two extremes (Lucas *et al.* 1978). The disk-shaped structure is also supported by a recent map of the molecular cloud in the NH_3 lines (Little and Macdonald 1978) and by the non-spherical distribution of absorbing material pointed out by Sibille *et al.* (1975). Located at the

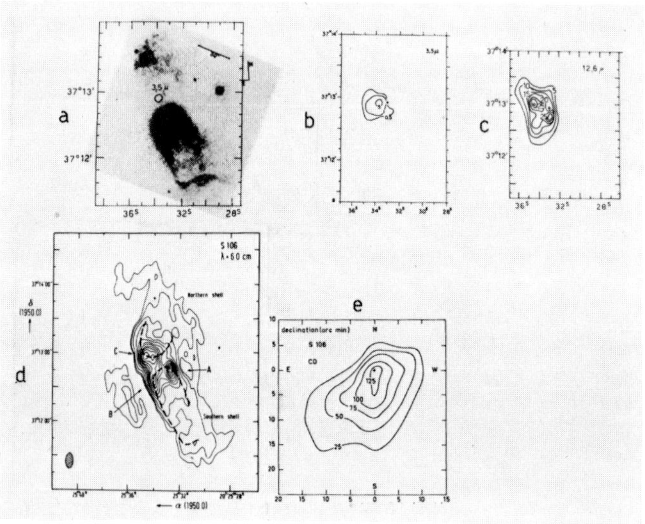

Figure 4: The S106 Region. a) Hα photograph from Maucherat (1975). b) 3.5 μm map from Pipher *et al.* (1976). c) 12.6 μm map from Pipher *et al.* (1976). d) 5 GHz map from Israel and Felli (1978). e) CO map from Lucas *et al.* (1978) (not in scale). The velocity varies from -3 km/sec in the NW part to +1 in the SE part.

centre of the two shell-like nebulosities is a strong infrared (1.25-3.4 μm) source, which is believed to be the main source of excitation of the nebula (Sibille *et al.* 1975). Recent 0.7-0.9 μm photographs of the region have indicated that this is the only point source in the area and that it is stellar-like and not a protocluster (Eiroa 1978). What is interesting in relation to the present discussion is that, although this strong near IR source is located at the centre of the radio structures, there is not any particular radio feature associated with it. On the contrary, the components A, B and C of the radio map, show a very close correlation with the 12.6 μm structures, indicating the coexistence of dust and ionized gas in the same regions. One possible solution is therefore that the near IR component is indeed the only source of excitation for the entire complex. Its Ly-c radiation is absorbed by dust in the plane of the disk and re-emitted by dust heated at a temperature of ≃ 2000 K. Part of the Ly-c flux is, however, capable to leak out in the two opposite directions perpendicular to the disk, where the bright components and the shell like structures are created. This picture is consistent with the difference in radial optical velocity in the northern and the southern part of the nebula (Maucherat 1975) and with the difference in obscuration if the "disk-like" molecular cloud is slightly tilted with respect to the line of sight.

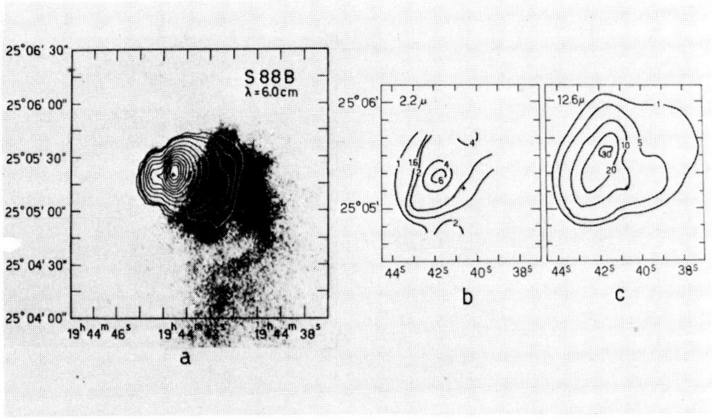

Figure 5: The S88B Region. a) Overlay of the H_α photograph (Deharveng 1978) with the WSRT 5 GHz map (Felli *et al.* 1978c). b) 2.2 μm map from Pipher *et al.* (1977). c) 12.6 μm map from Pipher *et al.* (1977).

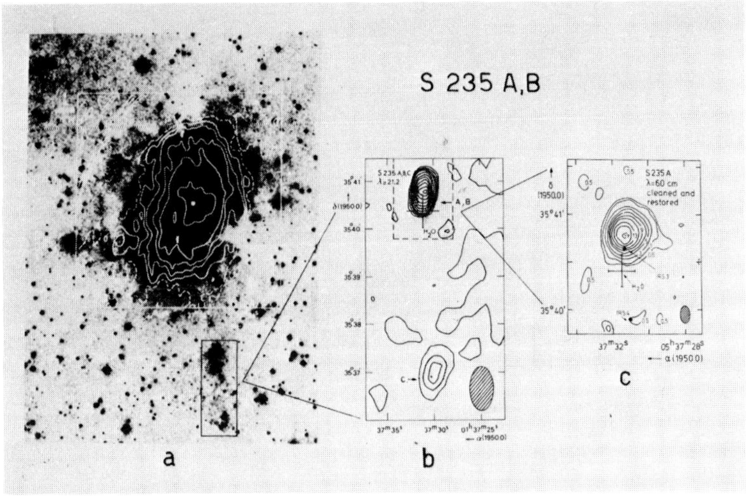

Figure 6: The S235 A, B Region. a) Overlay of the WSRT 1415 MHz map with the red Palomar Sky Survey print. b) WSRT 1415 MHz map of the S235 A, B, C region. c) WSRT 5 GHz map of the S235 A, B region. All maps from Israel and Felli (1978).

The non-uniform distribution of dust around exciting stars has also been found in other cases, by comparying radio, infrared and optical peak positions. For instance in K3-50 (Wynn-Williams *et al.* 1977), the radio (15 GHz) and middle IR (10 μm) peaks coincide but are displaced by several arc seconds from the optical nebulosity, while the 2.2 μm and 1.65 μm peaks lie between the radio and optical ones. This can be explained if the H II region is near the edge of a molecular cloud and if there are very strong extinction gradients across the ionization front that can shift towards regions of lower extinction the apparent peak at shorter wavelengths. A similar offset of radio, optical and IR peaks - in this case in a different order - occurs in S88 B. The radio map at 5 GHz (Felli *et al.* 1978c) is shown in figure 5, overlaid on the H_α photograph of Deharveng (1978) and compared with the 12.6 and 2.2 μm maps of Pipher *et al.* (1976). Also in this case, radio and optical peaks are clearly offset but here the 12.6 μm peak is located at the intermediate position and the 2.2 μm peak coincides with the brightest part of the radiomap, where possibly an even more compact source is present.

In short, all these examples point out that there are not only dust absorption effects in dense H II regions, but also indicate that the distribution of the dust and molecular gas around exciting stars is far from uniform and spherical. I shall discuss in more detail further on this "blister-like" configuration.

2.3 Free-free self absorption effects

The optical depth of an H II region with uniform density n_e and geometrical thickness d is $\tau_{ff}\ \tau\nu^{-2.1} \cdot EM$ where

$EM = \int_0^d n_e^2 dx$ is the emission measure. The radio flux density is directly proportional to n_e^2 only in the approximation $\tau_{ff} << 1$. For $\tau_{ff} > 1$ only a fraction $(e^{-\tau_{ff}} - 1)/\tau_{ff}$ of the radio emission is observed. In terms of EM this means that the condition:

$$EM << 3.05\ 10^6 \left[\frac{\nu}{GHz}\right]^{2.1} \tag{2}$$

must be satisfied. For instance, for ν = 5 GHz, EM has to be less than $9\ 10^7$ pc cm^{-6}. Taking a typical size of 1 pc, this implies $n_e < 10^4 cm^{-3}$.

As clear examples of this effect, W3 OH (Harten 1976; Harris and Scott 1976) and component G0.7-0.0 in Sagittarius B2 (Hobbs *et al.* 1971; Felli *et al.* 1974; Thum *et al.* 1978) can be quoted. More recently other very dense components have been found by measuring their continuum radio spectrum: the OH/H_2O source G 12.2-0.1 (Shaver and Danks 1978) and component E in M17 (Montgomery *et al.* 1971; Fukui and Iguchi 1977). The latter, although M17 is a fairly well studied object, was not found until very high frequency (87 GHz) observations were made. The turn-over frequency corresponds to EM ≈ 7×10^{10} pc cm^{-6}. It is coincident with a 21μm component (Lemke and Low 1972) and is located in a bright optical area, far from the ionization front detected at lower frequencies (Matthews, Harten and Goss 1978). A less clear, but possibly even denser case might be S235B. This is a small patch of nebulosity visible in H_α(Gluskov *et al.* 1975) and in the near IR (Sibille *et al.* 1976). No radio emission was detected by Israel and Felli (1978) down to the level of 5mJy at 5 GHz, as indicated in figure 6. Using the densities derived from optical observations, this globule would be optically thick also at 5 GHz (as M17E) and therefore undetectable at this frequency. Clearly higher frequency observations are necessary in this case to confirm this hypothesis. As pointed out by the above examples, the extreme case of highly self-absorbed components can be well studied when these are isolated strong sources. It is more difficult to reveal them when the region is composed of many clumps of different τ and the overall source size is of the order of the resolution of the telescope, so that the various components cannot be individually separated. In these cases only a careful analysis of the continuum radio spectrum will be able to determine the relative importance of high density clumps to lower density structures.

2.4 The effects of density boundness

A uniform H II region centered around an exciting star will be density bounded if the electron density is less than that required to balance the ionizations from stellar Ly-c photons. With a non-symmetrical density distribution, an H II region may be density bounded over a limited solid angle, while being ionization-bounded over the remaining part.

How can it be estimated whether an H II region is density bounded or not?

I) When both the exciting star and the radio optical nebula are clearly visible the best way is to examine the relative position of the two. Offsets, denouncing irregular distribution of gas around the star, partial shell structures and filamentary nebulae will be indications of this effect. There are many example of this configuration, see e.g. S188, S298 (Israel and Felli 1976). Quite often this is connected with star loosing mass, such as in NGC 6888 (Wendker *et al.* 1975).

II) A second possibility, when there are no outstanding geometrical effects, is to use (1) itself, when on other grounds it is known that effects a) and c) are not important. For instance it is interesting to note that the majority of the regions not included by Georgelin *et al.* (1975) in their calibration sample, fall in the density bounded part of the Log $(N_L^*/\pi F_\nu)$ - S_p diagram.

III) A third, more complex, case occurs, when the nebula shows a clear ionization front and the exciting star is located at the side of a molecular cloud. While the H II region will be ionization bounded on the side facing the molecular cloud, the same may not apply in the opposite direction, if the flow of gas produces a steep decrease in the radial electron density distribution.

The formation of an H II region at the edge of a molecular cloud has been followed in its dynamical evolution by Tenorio-Tagle (1978). He starts with an ionization bounded nebula inside and close to the outer edge of a molecular cloud. As soon as the ionization front reaches the surface of the molecular cloud, the ionization front will speed up in the diffuse medium surrounding the molecular cloud, which will be followed by the expansion of the H II gas, leaking through the whole of the cavity ("blister") at high speed, owing to the strong difference in pressure. Clearly in this case, the H II region will be density bounded (by definition of blister) in the corresponding solid angle and ionization bounded within the cavity.

More generally for an H II region with constant density up to a certain radius r_d and then with a power-law radial density distribution ($n_e \propto r^{-2}$, as can be expected in a radial steady flow at constant velocity) it can be easily shown (Felli and Panagia 1978) that there is a critical value $r_{crit} = r_o \cdot 0.69$, where r_o is the Stromgren radius in the case of uniform density distribution, such that when $r_d > r_{crit}$ the region is ionization bounded and when $r_d < r_{crit}$ the region becomes density bounded. It is also worth remembering that in these cases the spectrum will have a $\nu^{-0.6}$ behaviour over a certain frequency interal and this will be increased in relation to the range of radii over which the $n_e \propto r^{-2}$ distribution holds. (Panagia and Felli 1975).

After the first suggestion that a blister type configuration might be present in Orion, many more H II regions have been found to be associated to molecular clouds and to have a similar form.

The general aspect of H II regions with regard to molecular clouds will be discussed in section 3. Now I want to consider the improvements to the application of (1) when density boundness effects are present.

One correction is naturally to restrict the use of (1) to solid angles within which the H II region can be safely considered ionization bounded.

I) A first approximation is to assume the projected distance between the source of ionization (visible star or near infrared point source) and the ionization bounded part of the nebula as the true one. In this way the projected solid angle can be estimated and the value of N_L^* scaled accordingly to be compared with N_L^R of the ionization bounded part. This has been applied, for instance, to S252, a typical example of an H II region bounded by a molecular cloud on one side and expanding on the other. In this case a good match was found between N_L^* and N_L^R, both for the ionization front on the side of HD 42088 (component F in fig. 7) and for the extended low surface brightness region as a whole. For the more distant isolated components A, B and C, the Ly-c photon flux from HD 42088 in the corresponding solid angles were less than the observed N_L^R, and a local source of ionization was required for each of these components (Felli *et al.* 1977). A somewhat similar solid-angle analysis was made of the radio emission in W3A (Harris and Wynn-Williams

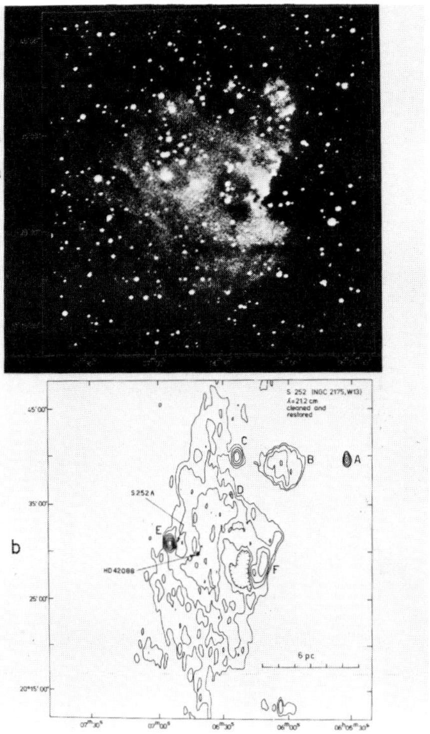

Figure 7: The S252 Region. a) Hα photograph from Lortet-Zuckerman. b) WSRT 1415 MHz map from Felli *et al.* (1977).

(Fig. 7a was first published in 'Lecture Notes in Physics', Vol. 42, p. 183; reprinted by permission of Springer-Verlag.)

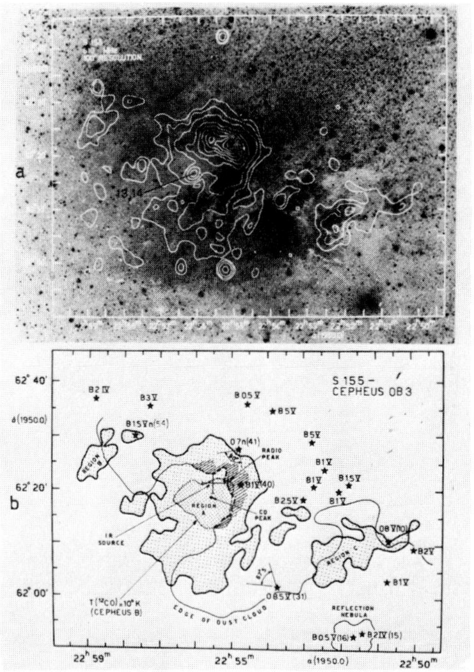

Figure 8: The S155 Nebula-Cepheus OB 3 association. a) Overlay of the WSRT 610 MHz map (resolution (112"x130") on the red Palomar Sky Survey print. b) Sketch of the main features in the S155 Nebula-Cepheus OB 3 association. All figures from Felli *et al.* (1978d).

1976). The radio emission was divided into equal cones centered on IRS2 and it was found that the geometry of the observed radio structure (in this case ionization bounded) was consistent with the presence of two displaced exciting stars, IRS2 and IRS2a.

II) The projected distance is clearly the lowest possible value to the true distance between the exciting star and the nebulosity. The correction will, therefore, tend to give overestimated values of N_L^*. A further refinement would be to determine the true three-dimensional configuration between star and nebulosity. Unfortunately this is hardly ever possible. However, if the spectral type of the exciting star is the best known parameter in the analysis and an offset is suggested by the configuration of the region, the whole argument can be reversed and from the observed value of N_L^R the solid angle subtended ψ can be derived with the relation.

$$\psi = N_L^R / N_L^* \qquad (3)$$

Such an argument has been applied to the S 155 nebula (Felli *et al.* 1978d). In this case the relative positions of the dark cloud-CO molecular bright peak with respect to the radio and optical maps (see fig. 8) suggest that the most important source of excitation (the O7n star HD217086 of the Cepheus OB 3 association) might be located on the far side of a molecular cloud. By use of eq. (3) the solid angle ψ and the distance of the star from the nebula were derived, after correcting N_L^R for the presence of dust absorption. The three-dimensional configuration is in agreement with the stronger reddening found for this star with respect to other nearby association members. With the derived distance, the stellar radiation from HD 217086 is a sufficient source of heating of the molecular cloud in order to explain the observed $C^{12}O$ brightness temperature.

2.5. Ionization balance over different scale sizes

As discussed in section 2.3 and 2.4 H II regions are known to have components with different densities; in particular they may be formed for the greatest part (in mass and volume) by very diffuse low density extended structures. The problem of detecting regions with very low EM may then become instrumental when the limits of sensitivity of the instrument are reached. Aperture synthesis instruments are limited to much higher EM ($\geq 10^4$ pc cm^{-6}) than single beam telescopes and therefore will tend to select intrinsically denser components. The single beam telescopes, which can reach the lowest EM, are limited to 100-300 pc cm^{-6}; e.g. a sphere of H II completely filling the beam of the NRAO 300" at a distance of 3 kpc (linear size 10 pc) will not be detectable if $n_e < 3$ cm^{-3}. Therefore, in the power-law electron density distributions considered in section 2.4 the boundary to the H II region may be set by the sensitivity of the instrument rather than by the intrinsic frontiers of the region itself. The distinction of H II regions into classes according to size and density is often used and originates from a possible evolutionary sequence for these objects. However, in the use of (1) one cannot limit oneself only to one class of objects, for instance to the more compact components. In many cases, these may emit only a small fraction of the total radio flux which is instead contained in lower density structures. Consequently, any speculation regarding the spectral type of the exciting star, based solely on the radio observation at high resolution, may be unsupported if lower resolution data are not examined to rule out the possible existence of extended envelopes.

When several exciting stars are present within the same H II regions, two different types of ionization balance can be made:

I) The total ionization balance, i.e. the comparison of the integrated radio flux from the entire region with the sum of the N_L^* values.

II) The study of selected radio structures observed with high resolution, taking into account the relative positions and subtended solid angles.

Both type of analysis were applied to S155, one of which has already been discussed in section 2.4. For the total ionization balance the sum of N_L^* values is about a factor 8 greater than the observed radio flux density from the same area. Using the volume occupied by the OB association, an upper limit of 5 cm^{-3} can be put to the electron density in the same region. Such a low density H II region would therefore still be fully ionized by the available stellar Ly-c photons, but, at the same time would be below the detection threshold.

These rarefield H II regions are the latest evolutionary stages in which they can be still seen as independent entities. The λ Orionis region, with a mean density of 2 cm^{-3} and an extension of 440 pc (Reich, 1978) and possibly the northern rim in the Monoceros Loop (Kirshner *et al.* 1978) might be examples. They can no longer be studied as single objects beyond this stage; their cumulative effect in the form of a diffuse ionized interstellar medium is all that remains to be studied.

2.6. The diffuse ionized interstellar edium

Of 2883 OB stars in the solar neighbourhood, about 50 per cent are not associated with visible H II regions, i.e. with EM > 50 pc cm^{-6} (Torres-Peimbert *et al.* 1974). The integrated spectrum emitted by these "naked" stars is similar to that of an O7 and this radiation may be a very important source for the ionization of the interstellar medium. Mezger and Smith (1975) brought this percentage to an even higher value of 80-90 per cent when considering OB stars in the whole Galaxy embedded in compact H II regions (defining compact H II regions as those in which most of the stellar Ly-c photons get absorbed within one pc of the exciting star).

The presence of diffuse ionized gas in the plane of the Galaxy is now well established on the basis of several independent observations:

a) The thermal component of the radio continuum galactic background (Westerhout 1958).

b) The radio recombination lines from regions free of discrete H II regions (Gordon and Gottesman 1971; Hart and Pedlar 1976; Lockman 1976).

c) The radio recombination lines observed in the direction of SNR's and attributed to diffuse H II regions along the line of sight (Downes and Wilson 1974).

d) The absorption in the low frequency part of the radio continuum spectrum of galactic nonthermal sources, also attributed to diffuse ionized gas along the line of sight (Dulk and Slee 1972).

e) The pulsar dispersion measurement.

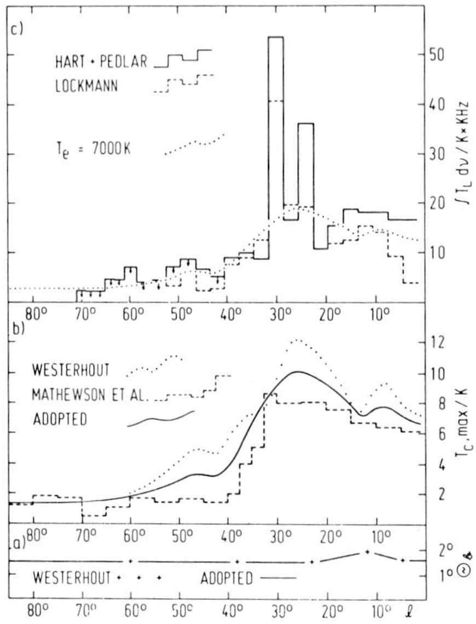

Figure 9: The emission of the Galactic low-density H II region. a) HPW in galactic latitude of the free-free continuum emitted at $\nu = 1.39$ GHz by the galactic ELD H II region. b) Peak brightness temperature of the free-free continuum emitted at $\nu = 1.39$ GHz by the galactic ELD H II region. c) Integrated H 166α recombination line intensity ($\nu = 1.425$ GHz) emitted by both the radio H II regions and galactic ELD H II region. The curve scaled from the adopted free-free continuum emission for Te= 7000 K relates only to the latter component. From Mezger (1978).

A connection between this low density extended H II region and the OB stars located outside distinct H II regions has been done by Mezger (1978) by means of an ionization balance over a large scale.

Using the LTE relationship between $\int T_L d\nu$, the integral of the radio recombination line profile (H 166α), and T_c, the continuum brightness temperature of the thermal background emission, he related the longitudinal distribution of $\int T_L d\nu$ with the analogous distribution of T_c (see fig. 9). The best fit temperature found was Te ≈ 7000 K. The integrated flux density in this extended low density H II region was found to be seven times greater than that emitted by giant H II regions. With a simple galactic model of concentric rings and assuming that the Ly-c photon flux in each ring is a function only of the distance from the galactic centre, the values of $N_L^R(r_i)$ were found by fitting the observed (T_c-galactic longitude) plot. The values of N_L^R found are limited to a distance range from 4 kpc to 13 kpc and have a maximum at r_i = 8 kpc.

The values of $N_L^R(r_i)$ were then corrected for absorption by dust in the galactic plane and density boundness effects in the direction perpendicular to the galactic plane. In this way, N_L^*, the flux of required stellar Ly-c photons was derived. This is 2.2 10^{53} photons sec^{-1} $\pm$ 20-30 percent.

The corrected values of $N_L^*(r_i)$ can now be directly compared with the flux of Ly-c photons produced by OB stars not associated with H II regions, (Torres-Peimbert *et al.* 1974). The agreement for r_i = 9-11 kpc (the solar neighborhood) is satisfactory, and implies that OB stars outside H II regions are indeed responsible for the ionization of the extended low density H II region. However, the corrected $N_L(r_i)$ increases from r_i = 10 kpc to r_i = 5 kpc, by approcimately a factor five. This increase of production of Ly-c photons can only be derived from an analysis of radio data. An extrapolation of the optical results to the inner part of the Galaxy would strongly underestimate the number of Ly-c photons.

As a product of this analysis, from the ratio of the total Ly-c photons emitted by stars outside H II regions (ΣN_L^* = 2.2 10^{53} s^{-1}) to that emitted by stars within giant H II regions (ΣN_{Lgiant}^* = 0.5 $10^{53} s^{-1}$), Mezger (1978) derives the lifetime of a giant H II region,

$$t(\text{giant H II}) = \frac{\Sigma N^*_{L\,giant}}{(\Sigma N_L^* + \Sigma N^*_{Lgiant})} < T_{MS} > = 8.1\ 10^5\ \text{yr}$$

where $<T_{MS}>$ is the mean main sequence lifetime of an OB star.

3. H II REGIONS AND STAR FORMATION

H II regions are known to represent early manifestations (although not the earliest) accompanying the formation of hot stars and hence their properties can be used to test the proposed mechanism of star formation. Woodward (1978) has summarized four possible ways of suddenly increasing the external pressure around dense cool molecular clouds in a stage of quasi-equilibrium and to force them to collapse, namely: a) the flow through a spiral arm of a density wave; b) the interaction with the strong stellar winds; c) the expansion of a supernova remnant and d) the expansion of an H II region. Of these, the last three mechanisms are of a chain type, i.e. once a star is formed, the dynamical effects produced by this may start up the formation of subsequent generations of stars and thus iterate the process as long as there is material to convert into stars.

To my knowledge case a) does not have clear individual examples, and only IRSS in IC 1805 has been suggested for case b), (Vallé and Hughes 1978).

The supernova cascade process for producing stellar associations has been discussed by Ögelman and Maran (1976). The examples most frequently quoted in support of this mechanism are the Perseus OB 2 association (Sancisi *et al.* 1974), the CO cloud connected to the W 44 SNR (Dickel *et al.* 1976), the Canis Majoris R1 association (Herbst and Assousa 1977) and the Ori-Gem Loop (Berkhuijsen 1974). However, in a more careful analysis, all the above examples with the possible exception of the last one refer more specifically to the interaction of the supernova shell with the surrounding medium (H I, CO) rather than proving induced star formation.

Also, the distinction between the observed effects produced by the expansion of a SNR and those of the expansion of a very diffuse extended H II region ionized by an OB association may be very difficult and not unambiguous, due to the close resemblance of the faint ring structures produced in the two cases and to the similar dynamical effects (Lasker 1967). For instance, the two alternative possibilities c) and d) have been proposed to explain the ring of neutral hydrogen expanding around the Cepheus OB 3 association (Assousa *et al.* 1977; Felli *et al.* 1978 d). It can be said in favour of the H II region hypothesis that the OB stars are directly observable and an estimate of their contribution to the energetics of the H I shell gives a result equal to the observed one, while the presence of a SNR is not supported by direct independent evidence and must be postulated.

In the model for sequential formation of subgroups in OB associations proposed by Elmegreen and Lada (1977), a shock front and an ionization front propagate at the edge of a molecular cloud, driven by the Ly-c radiation of an earlier generation of OB stars. After about 2×10^{6} yr at the interface between the H II region and the molecular cloud the densities are increased to the point that conditions for gravitational collapse are reached. A new generation of hot stars will then form, surrounded by compact H II regions. These in turn will reach the edge of the molecular cloud, disperse the H II gas into the extended H II region on the side of the molecular cloud and send a shock-ionization front into the molecular cloud, thus repeating the process.

It is not clear at present whether one requires pre-existing dense globules in the neutral gas, which will be forced to collapse by the increase in pressure at the shock front, or an increase in density of the shock which can itself create collapse conditions. In the CO hot spot at the edge of the S 155 nebula, Cepheus OB 3 association, where we suggest that this process might be operating, the observed density and mass indicate that the molecular cloud should collapse on a time scale of 3×10^{5} yr. This is short compared to the estimated age of the association (4×10^{6} yr) and may suggest that the present density in the cloud is the result of a recent compression and not preexisting to the stars of the association.

The dynamic of the H II gas in the later stages of this model and those of the blister model discussed in section 2.4 are fairly similar, the main difference being that the blister model requires the presence of a high density well defined cavity. This blister configuration was first proposed for Orion A by Zuckermann (1973) and Balick *et al.* (1974) and was later on found to be applicable to many more H II regions (see e.g. Deharveng *et al.* 1976; Israel 1977). The net result in terms of H II region evolution is that, after the ionized gas breaks though the molecular cloud, we will not observe the standard expansion of a spherical, uniform density H II regions, such as that studied by Mathews (1969), but rather the flow of ionized gas streaming off the side of a molecular cloud.

There are several examples in which star formation seems to follow this pattern. Of the examples quoted by Elmegreen and Lada (1977),

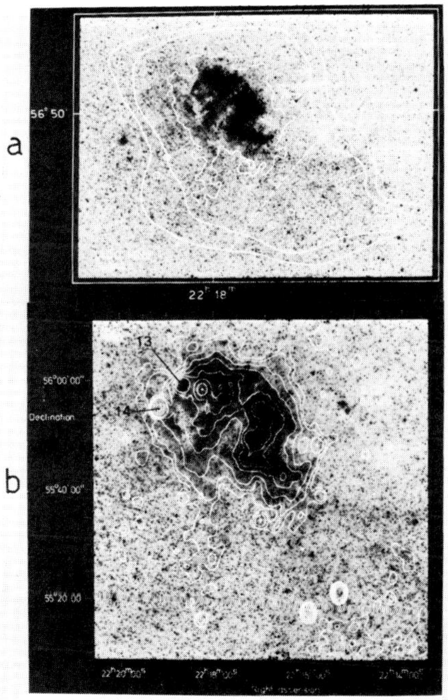

Figure 10: The S132 region. a) The large scale structure of the radio emission at 1.4 GHz associated with S132, overlayed on the red Palomar Sky Survey print. The two outer contours are the 1.2 k and 1.8 k contours of the 1.4 GHz map of Felli and Churchwell (1972). For reference, the 5 mJy contour of the 0.610 GHz map is also included. b) WSRT 0.610 GHz map of S132 with a resolution (112"x120") overlayed on the red Palomar Sky Survey print. All figures from Harten *et al.* (1978).

namely NGC 7538, M8 and M17, the last one seems to be the most clear (Lada 1976; Matthews *et al.* 1978). Another possible evidence of star formation which took place at different positions and moved in the direction of a molecular cloud may be S132 (Harten *et al.* 1978), see figure 10. The S-W part of this large H II region is very diffuse and without bright structures. The exciting stars are all visible, since a good ionization balance can be performed. Hence this part can be considered as the one in which star formation first occurred. Moving to the N-E direction we find brighter optical and radio structures mixed with patches of obscuring material. The stars found in this area are not enough to balance the observed ionization and hence some of them may still be hidden behind blobs of neutral material, a situation reminiscent of the phase in which stars begin to disperse the material out of which they were formed. At the extreme N-E side of the nebulosity there is an ionization front at the edge of an obscuring cloud, which may represent the site in which further star formation will eventually occur.

One of the main results of recent molecular (CO, H_2CO, NH_3) observations in the direction of H II regions (in the majority of the cases diffuse ones) is the establishment of a clear correlation between H II regions and molecular clouds, i.e. H II regions are, in the majority of the cases, found to be associated with molecular clouds. To what extent the reverse is also true or, alternatively, what the selection effects are of this relationship is not clear at this stage as pointed out by Harris (1978) and only unbiased searches will be able to provide us with an answer. In an attempt to study in detail this relationship, i.e. to determine whether blobs of H II and molecular gas are distributed at random throughout a spherical volume, or whether they are instead systematically correlated in the sense of H II regions always occurring at the side of a molecular cloud,Israel (1978) has considered a large sample of H II regions (60 in all) for which high resolution continuum radio observations, recombination lines (optical or radio) and CO observations are available. The aspects which Israel studied are:

a) the relative positions of H II and CO peaks;
b) the morphology of the radio continuum emission;
c) the relative radial velocities of H II regions and CO clouds.

The ratio of the distance between the brightest continuum radio peak and the brightest CO peak to the radius of the CO cloud has a mean value and a distribution which is consistent with that expected in the case of H II regions on the surface of a spherical cloud. Core envelope structures and curved ridges or incomplete shells in the radio maps (fig. 11) are suggestive of ionization fronts and of ionized gas streaming from a molecular cloud. These characteristics are more evident in the denser structures and are present in about half of the entire sample. I would add to this that structures of a similar shape have a wide range of linear sizes from less than 0.1 pc to 3 pc. This could imply that this configuration is present over a relatively long phase of life of an H II region and is not a transient stage.

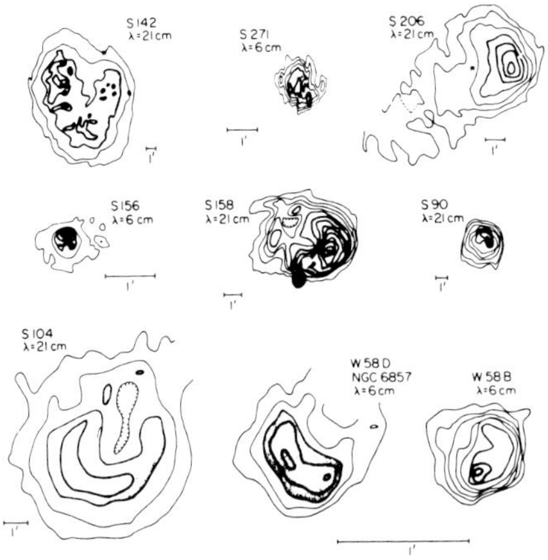

Figure 11. Radio structures of several H II regions. S156, W58 B, W58 D and S271 are on a linear scale three times larger than that of S90, S158 and S206; S142 is on a linear scale two times smaller. From Israel (1978).

The behaviour of the (V_{HII}-V_{CO}) distribution is the effect which most of all speaks in favour of a flow of H II gas from the surface of a molecular cloud. In optically visible regions the distribution (fig. 12) shows an offset from a mean zero value of -3.4 ± 0.4 km/sec. The clearly indicates that when the H II region is on the side of the molecular cloud facing the observer, the H II gas is moving in this direction, streaming off the molecular cloud at an average escape velocity of - 3 km/sec.

The distribution of (V_{HII} -V_{CO}) for the opposite case, i.e. the optically obscured regions, does not show a symmetrical effect, i.e., a ΔV positive, but a mean zero value (fig. 12). This distribution can still be explained in the frame of the blister model if the obscuration is produced not only by local matter but also by matter distributed along the line of sight.

One of the sources in Israel's sample is S155 to which we have already referred in several instances and which, according to our geometric model, should be placed mainly on the far side of the CO cloud. The observed (V_{HII}-V_{VO}) is -2.8 ± 2.5 to be compared with a positive value expected from the geometric model. However, the observed ΔV does not contradict the model since the H II velocity is from Hα measurements and the optically visible part of the ionized gas flows in the direction of the observer. Observations of radio recombination lines

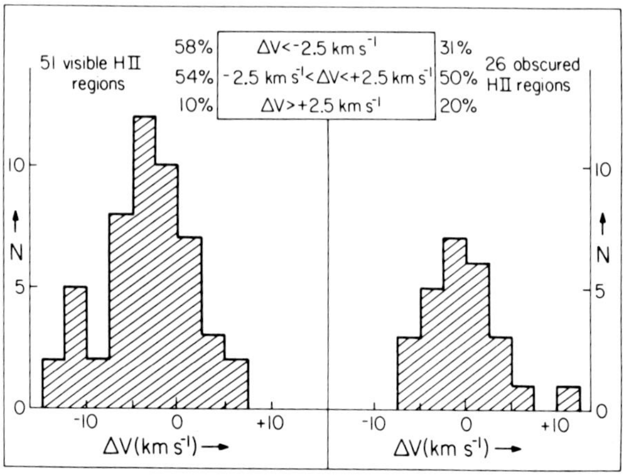

Figure 12. Distribution of the velocity differences V(H II-CO) for visible and for obscured H II regions. From Israel (1978).

are in progress to clarify the geometrical structure in this particular case where there are other independent indications of the three-dimensional configuration.

The clearest evidence that star formation might take place at the edge of an extended H II region expanding into a molecular cloud is the W3 complex, located on the west side of the bright nebulosity IC 1975. Incidentally, we note that this region - being located in a heavily obscured area - has been, historically, the prototype of the model in which star birth occurs at the centre of a molecular cloud (see e.g. Schraml and Mezger 1963).

If the relative position of W3 to IC 1975 is taken as the proof of the validity of the Elmegreen and Lada model one is led to generalize that high density thermal radio components should always be confined to the ionization fronts separating diffuse H II regions from molecular clouds.

Before examining the observations made in this respect, it is useful to recall - as pointed out by Habing (1975) - that two different types of high density thermal radio components may exist, both having similar observable properties but each related to intrinsically different configurations and subsequent evolutionary stages. One is the compact H II region outlined by Schraml and Mezger (1963) representing the high density Strömgren sphere, completely ionization bounded, surrounding an early-type star. This phase might be present also in the blister model before the edge of the molecular cloud is reached. The other type of high density thermal sources are the brightest parts of the ionization fronts. Being close to the molecular cloud, electron densities up to 10^3-10^4 cm^3 can be reached (especially in favorable geometrical configurations), while the exciting star is well outside the "cavity". The two configurations can barely be distinguished using radio continuum data only. Higher densities are in favour of the "compact H II regions", but are not a sufficient criterion. To be able to distinguish between the two, one needs either optical observations to reveal if stars are directly associated with the suspected components, or near-IR observations to determine the temperature of dust, which is expected to be higher the nearer it is to the ionizing star.

Examples of the second type (although not the extreme cases could be the globule no. 9 in S155 (fig. 8) or no. 13 in S132 (fig. 10).

Now that the presence of H II regions on the side of molecular clouds has been proved in several cases, it is interesting to answer the following questions: a) how frequent is this configuration and b) can traces of recent star formation be found.

In order to give a qualitative answer to the first question I have considered a large sample of visible H II regions (including Israel's sample) and have attempted to divide them into two classes: regular and irregular. I have considered as regular those H II regions with an approximately spherical shape, a central exciting star, no obscuration effects due to local dust, no molecular clouds, and no ionization fronts or bright sharp rims bounded by obscuration fronts. The distinction is therefore purely morphological and not based on homogeneous data. The percentages found are about 15 percent for the regular and 85 percent for the irregular. From this limited analysis I can say that in a minority of cases the peculiar location of the H II gas on the side of a molecular cloud can be excluded, while there is a great number of cases in which this possibility is open. It is also worth pointing out that in the regular H II regions there are no bright components, all of them being faint and diffuse.

The abundance of rim structures in H II regions is by no means a new phenomenon. Bright rims had been studied, for instance, by Pottasch (1956) who revealed a relationship between the shape of the rim and the distance to the exciting star. Figure 2 of his paper is an excellent sketch representation of this configuration. What we now know is that fairly often these rims are on the side of massive molecular clouds.

Given the possible high frequency of this configuration and the scarce dependence on time of the electron density distribution in the diffuse H II gas, it is clear that estimates of the age of an H II region from its mean density (which is a valid criterion in the regular ones) generally have little meaning.

Are the regular H II regions simply the final stages of irregular ones, where all the material has been ionized and dispersed into the interstellar medium? Or do they represent a distinct class of H II regions with different initial triggering of star formation and different evolution? In order to be able to provide answers and not just questions it is necessary to carry out more work in this respect.

Examples of star formation taking place in the ionization front (of the W3 type) are more difficult to find. In W40 and W48 the continuum radio peaks (Felli *et al.* 1974) are coincident with the 2.2 μm peaks and both are displaced from the CO peaks (Zeilik and Lada 1978). This is interpreted as a typical blister type object, but no traces of young stars have been found at 2.2 μm at the interface and/or close to the CO peaks (Zeilik and Lada 1978).

To find out how often compact H II regions can be found in the proximity of ionization fronts a large sample of H II regions, known to be thermal radio sources from previous low resolution radio observations, have been searched at high frequency (5 GHz) and high resolution (∿6"). Small-diameter weak radio structures could in fact have been smoothed out and lost in lower resolution radio observations. The sample contained 77 H II regions. In 34 of them a total of 47 small-diameter components were detected (Felli *et al.* 1978b). The individual analysis of each case is still in progress, but from a preliminary inspection of the results and also using the results of other synthesis observations (see e.g. Israel 1978) we can say that in only very few cases the detected components can be unambiguously explained in terms of compact components associated with recently formed stars and located close to the molecular cloud. A somewhat complementary study was carried out by Gilmore (1978) who searched compact H II regions in many dark clouds. Also in his survey no hidden compact H II regions imbedded inside dark clouds were detected, whereas in several cases lower density H II regions were found to be located at the edges of molecular clouds.

There is the possibility that the very small percentage of known recently formed stars may be produced by an instrumental bias. In fact, the very high density compact components searched will almost certainly suffer from strong self-absorbtion effects and dust absorption, and thus 5 GHz radio observations may not be the best way to search them (higher frequency radio observations or 2-10 μm IR observations would be better).

However, the radio continuum observations indicate a very short life-time of the compact components with respect to the life-time of the diffuse H II regions. This is in agreement with the long time scales for the formation of instability conditions at the ionization front, ($2\ 10^6$ years, Elmegren and Lada 1977) and the short time scales for the evolution of compact H II regions (10^4 years, see e.g. Mezger 1978).

In short, there is increasing evidence demonstrating that H II regions cannot be looked upon any more only in the classical frame of a uniform spherical cloud of ionized gas around a hot star. H II regions are found at the edges of molecular clouds and may provide a powerful mechanism to compress the gas and to trigger star formation.

What do we know about the very early phases of star formation? Very little. In this context it is worth mentioning, as a simple curiosity, that in five cases non-thermal unresolved components have been found at the position where one would expect star formation to take place: component D in S252, figure 7; components no:s 13 and 14 in S155, figure 8; component no. 14 in S132; figure 10; one component in W1, Harten (private communication) and 4C67.34 in NGC 7023 (Pankonin and and Walmsley 1978). Of course the easiest explanation is that they are extragalactic background sources. But do we know enough about protostars to safely disregard non-thermal emission? What about the magnetic energy dissipated during the collapse, see e.g. Huges (1969)?

It is necessary to caryy out further work in the field of H II regions in order to answer our final questions as well as all the previous ones we have put forward.

I want to thank all the colleagues who sent me their results prior to publication. Stimulating discussion with N. Panagia and G. Tofani helped me constantly during the preparation of this paper.

REFERENCES

Assousa, G.E., Herst, W., Turner, K.C.: 1977, Astrophys. J. 218, L13.
Balick, B.: 1975, Astrophys. J. 201, 705.
Balick, B., Gammon, R.H., Hjellming, R.M.: 1974, *H II Regions and the Galactic Centre,* 8th ESLAB Symposium, p. 135, Ed. A.F.M. Moorwood, ESRO SP 105.
Berkhuijsen, E.M.: 1974, Astron. Astrophys. 35, 429.
Blair, G.N., Evans, N.J., VandenBout P.A., and Peters, W.L. III: 1978, Astrophys. J. 219. 900.
Brown, R.L., Lockman, F.J., Knapp, G.R.: 1978, Annu. Rev. Astron. Astrophys., in press.
Churchwell, E.: 1974, IAU Symp. 60, 195; Eds. F.J. Kerr and S.C. Simonson, III, D. Reidel Publ. Co., Dordrecht.
Churchwell, E., and Walmsley, C.M.: 1973, Astron. Astrophys. 23, 117.
Deharveng, L.: 1978, Submitted to Astron. Astrophys.
Deharveng, L.,Israel, F.P., Mancherat,M.: 1976, Astron.Astrophys. 48, 63.
Dickel, J.R., Dickel, H.R., Crutcher, R.M.: 1976, Publ. Astron. Soc. Pacific 88, 840.
Downes, D., and Wilson, T.L.: 1974, Astron. Astrophys. 34, 133.
Dulk, G.A., and Slee, O.B.: 1972, Australian J. Phys. 25, 429.
Eiroa, C.. 1978, private communication.
Elmegren, B.G., Lada, C.J.: 1977, Astrophys. J. 214, 725.
Felli, M., Churchwell, E.: 1972, Astron. Astrophys. Suppl. Ser. 5, 369.
Felli, M., Habing, H.J., and Israel, F.P.: 1977, Astron. Astrophys. 59,43.
Felli, M., Harten, R.H., Habing, H.J., and Israel, F.P.: 1978b,
Felli, M., Israel, F.P., and Blair, G.N.: 1978c, in preparation.
Felli, M., Natta, A., Panagia, N.: 1978a, in preparation.
Felli, M., Panagia, N.: 1978, in preparation.
Felli, M., Tofani, G., and D'Addario, L.R.: 1974, Astron. Astrophys.31,431.

Felli, M., Tofani, G., Harten, R.H., and Panagia, N.: 1978d, Astron. Astrophys., in press.
Fukui, Y., Iguchi, T.: 1977, Publ. Astron. Soc. Japan 29, 63.
Georgelin, Y.M., Lortet-Zuckermann, M.C., Monnet, G.: 1975, Astron. Astrophys. 42, 273.
Gilmore, W.S.: 1978, Ph. D. Thesis, University of Maryland.
Gluskov, Y.I., Denisyuk, E.K., Karyagina, S.V.: 1975, Astron. Astrophys. 39, 481.
Gordon, M.A., and Gottesman, S.T.: 1971, Astrophys. J. 168, 361.
Habing. H.J.: 1975, Lecture Notes in Physics no. 42, 156; Eds. T.L. Wilson and D. Lownes, Springer Verlag, Berlin.
Harris, S.: 1978, Paper presented at the XXII Herstmonceux Conference.
Harris, S., and Scott, P.F.: 1976, Mon. Not. R. Astron. Soc. 175, 371.
Harris, S., and Wynn-Williams, C.G.: 1976, Mon. Not. R. Astron. Soc. 174, 649.
Hart, L., and Pedlar, A.: 1976, Mon. Not. R. Astron. Soc. 176, 547.
Harten. R.H.: 1976, Astron. Astrophys. 46, 109.
Harten, R.H.: 1978, private communication.
Harten, R.H., Felli, M., and Tofani, G.: 1978, Astron. Astrophys., in press.
Harvey, P.M., Campbell, M.F., and Hoffmann, W.F.: 1978, Astrophys. J. 219, 891.
Herbst, W., Assousa, G.E.: 1977, Astrophys. J. 217, 473.
Hjellming, R.M.: 1968, Astrophys. J. 154, 533.
Hobbs, R.W., Modali, S.B., Maran, S.P.: 1971, Astrophys. J. 165, L87.
Hughes, V.A.: 1969, Nature 222, 733.
I.A.U. Symposium n. 75 *Star Formation,* 1977, Eds. T. de Jong and A. Maeder 1977, D. Reidel Publ. C., Dordrecht.
Israel, F.P.: 1977, Astron. Astrophys. 59, 27.
Israel, F.P.: 1978, Astron. and Astrophys., in press.
Israel, F.P., and Felli, M.: 1976, Astron. Astrophys. 50, 47.
Israel, F.P., and Felli, M.: 1978, Astron. Astrophys. 63, 325.
Israel, F.P., Habing, H.J., de Jong, T.: 1973, Astron. Astrophys. 27, 143.
Kazès, I., Le Squeren, A.M., Gadea, F.: 1975, Astron. Astrophys. 42, 9.
Kirshner, R.P., Gull, T.R., and Parker, R.A.R.: 1978, Astron. Astrophys., Suppl. Ser. 31, 261.
Lada, C.J.: 1976, Astrophys. J. Suppl. Ser. 32, 603.
Lasker, B.M.: 1967, Astrophys. J. 149, 23.
Lemke. D., Low, F.J.: 1972, Astrophys. J. 177, L53.
Little, L.T., Macdonald, G.H.: 1978, private communication.
Lockman, F.: 1976, Astrophys. J. 209, 429.
Lortet-Zuckermann, M.C.: 1975, Lecture notes in Physics 42, 183; eds T.L. Wilson and D. Downes, Springer-Verlag, Berlin.
Lucas, R., Le Squèren, A.M., Kazès, I., Encrenaz, P.J.: 1978, Astron. Astrophys. 66, 155.
Maucherat, A.J.: 1975, Astron. Astrophys. 45, 193.
Mathews, W.G.: 1969, Astrophys. J. 157, 583.
Matthews, H.E., Harten. R.H., Goss, W.M.: 1978, submitted to Astron. Astrophys.
Mezger, P.G.: 1978a, International School E. Majorana. 4th Course on *Infrared Astronomy*. Eds G. Setti and G. Fazio, Reidel Publ. Co.,

Dordrecht; in press.
Mezger, P.G.: 1978, submitted to Astron. Astrophys.
Mezger, P.G., and Smith, L.F.: 1975, in *Proc. of the third European Astronomy Meeting, Tbilisi;* 369; Ed. E.K. Kharadze, (Acad. Sci. Georgian SSR).
Mittelberg Symposium on *H II Regions and Related Topics:* 1975, Lecture Notes in Physics no. 42; Ed. T.L. Wilson and D. Downes, Springer-Verlag, Berlin.
Montgomery, J.W., Epstein, E.E., Olivier, J.P., Dworetsky, M.M. and Fogarty, W.G.: 1971, Astrophys. J. 167, 77.
Natta, A., and Panagia, N.: 1976, Astron. Astrophys. 32, 269.
Ögelman, H.B., Maran, S.P.: 1976, Astrophys. J. 209, 124.
Panagia, N.: 1973, Astron. J. 78, 929.
Panagia, N.: 1974a, Astrophys. J. 192, 221.
Panagia, N.: 1974b, private communication, unpublished.
Panagia, N.: 1978, International School E. Majorana, 4th Course on *Infrared Astronomy,* Eds. G. Setti and G. Fazio, Reidel Publ., Co., Dordrecht,
Panagia, N., and Felli, M.: 1975, Astron. Astrophys. 39, 1.
Pankonin, V., Walmsley, C.M.: 1978, Astron. Astrophys. 67, 129.
Petrosian, V., Silk, J., Field, G.B.: 1972, Astrophys. J. 177, L69.
Pipher, J.L., Sharpless, S., Savedoff, M.P., Krassner, J., Varlese, S., Soifer, B.T., and Zeilik, M.: 1977, Astron. Astrophys. 59, 215.
Pipher, J.L., Sharpless, S., Savedoff, M.P., Kerridge, S.J., Krassner, J., Schurmann, S., Soifer, B.T., and Merrill, K.M.: 1976, Astron. Astrophys. 51, 255.
Pottasch, S.: 1956, Bull. Astron. Inst. Netherlands 13, 77.
Reich, W.: 1978, Astron. Astrophys. 64, 407.
Rubin. R.H.: 1968, Astrophys. J. 154, 393.
Sancisi, R., Goss, W.M., Anderson, C., Johansson, L.E.B., Winnberg. A.: 1974, Astron. Astrophys. 35, 445.
Shaver, P.A., Danks, A.L.: 1978, Astron. Astrophys. 65, 323.
Schraml, J., and Mezger, P.G.: 1969, Astrophys. J. 156, 269.
Sibille, F., Bergeat, J., Lunel, M., and Kandel, L.: 1975, Astron. Astrophys. 40, 441.
Sibille, F., Lunel. M., Bergeat, J.: 1976, Astron. Astrophys. 47, 161.
Tenorio-Tagle, G.: 1978, Paper presented at the XXII Herstmonceux Conference.
Thum, C., Mezger, P.G., Pankonin, V., and Schraml. J.: 1978, Astron. Astrophys. 64, L17.
Torres-Peimbert, S., Lazcano-Araujo, A., Peimbert, M.: 1974, Astrophys. J. 191, 401.
Vallé, J.P., Hughes, V.A.: 1978, Astrophys. J. 223, 297.
Wendker, H.J., Smith, L.F., Israel, F.P., Habing, H.J., Dickel, H.R.: 1975, Astron. Astrophys. 42, 173.
Westerhout, G.: 1958, Bull. Astron. Inst. Netherlands 14, 215.
Woodward, P.R.: 1978, Ann. Rev. Astron. Astrophys., in press.
Wynn-Williams, C.G., Becklin, E.E., Mathews, K., Neugebauer, G., and Werner, M.W.: 1977, Mon. Not. R. Astron. Soc. 179, 255.
Wynn-Williams, C.G., Becklin, E.E., and Neugebauer, G.: 1972, Mon. Not. R. Astron. Soc. 160, 1.

Zeilik, M. II, Lada, C.J.: 1978, Astrophys. J. 222, 896.
Zuckermann, B.: 1973, Astrophys. J. 183, 836.

ACKNOWLEDGEMENT

Figures 3a, 6a, 8a, and 10a are © copyright by the National Geographic Society - Palomar Observatory Sky Survey and are reproduced by permission from the Hale Observatories.

DETECTION OF OPTICAL IMAGES

G. Sedmak
Osservatorio Astronomico di Trieste, Trieste, Italy

ABSTRACT

The detection of optical images is considered with emphasis given to the applications and technology oriented classifications and to the description of the available, basic state-of-the-art detectors. - The image detectors considered are examined with reference to their main astronomical applications. -

The trends of development for future image detectors of higher performance than those available now are considered and the most promising image detectors identified.

I. INTRODUCTION

The detection of optical images is the fundamental task of any research program concerned with optical astronomy.

An optical image is defined in general as the distribution of the source irradiance over the two-dimensional reference frame coincident to the observer and to the projected normal to the vector of propagation of the photon carrier.

The detection of an optical image is defined in general as the process that transfers the optical information of the image detected in the measuring time into an array of numbers to be used for the recovery of the scientific information contained in that image. The physical realizability of the detector implies that only a closed region of finite area of the image can be sampled by a finite spatial resolution in a finite measuring time and in the presence of a finite system noise.

The detection process is defined by the assignment of the detector transfer function. Consequently, the calibration of the detector must be considered as a fundamental component of the detection process.

Bengt E. Westerlund (ed.), Stars and Star Systems, 221–234.

The definitions given above can be used to organize a general classification of the basic systems available for the detection of optical images on the basis of the parameters that characterize the system transfer function. The basic systems can be described more advantageously on the basis of the technology used in their realization. The resulting scenary allows then to make an application-oriented comparison between the various systems and the identification of the trends of development of current technology in the detection of optical images.

2. CLASSIFICATION OF THE BASIC IMAGE DETECTORS

According to the approach given in Section 1 the classification of the basic image detectors must be based on the parameters that characterize the system transfer function with reference to the definition stated for an optical image.

The basic parameters to be considered are the following:

(1) *Geometrical parameters:* Image format and spatial resolution.
(2) *Photometric parameters:* Intensity dynamic range and noise.
(3) *Spectral parameters:* Spectral range and quantum efficiency.
(4) *Temporal parameters:* Time resolution and integration time.
(5) *Operational parameters:* Integration mode and number of individual detectors. System stability against time and environment and system reliability.

A separate technology-oriented classification is needed to accommodate the large variety of available image detectors into an application-oriented frame like that given above with a sufficient organization. A technology-oriented classification must be based on the basic operational features of the detectors: read-in, integration, and read-out mode.

According to this approach the technology oriented frame results as follows:

(a) *Read-in mode:* (1) Multielement detector systems.
(2) Multiplexed single detector systems.
(b) *Integration mode:* (1) Analog.
(2) Digital (photon counting).
(c) *Read-out mode:* (1) Parallel, destructive and non destructive.
(2) Multiplexed, destructive and non destructive.

The most commonly available basic image detectors can be classified in this frame as follows:

(A) *Analog and digital integrating multiplexed single detector systems:*
(1) Time multiplexing image scanners (optomechanical scanner, image dissector tube).

(2) Spatial frequency multiplexing image detectors (Hadamard transform imager).

(B) *Analog integrating multielement detector systems:*
(1) Photosensors arrays and intensified photosensors arrays in photon input array mode and:
(1.1) Non destructive sequential read out mode (Photographic film, CID solid state array; coupled to an image intensifier in intensified mode).
(1.2) Destructive sequential read-out mode (Vidicon tube, Reticon and CCD solid state arrays; coupled to an image intensifier in intensified mode).
(2) Intensified photosensors arrays in electron input array mode and:
(2.1) Non destructive sequential read-out mode (Electronography).
(2.2) Destructive sequential read-out mode (SEC and EBS tubes, self-scanned Digicon).

(C) *Digital integrating multielement detector systems:*
(1) Photon counting imager with buffer memory and sequential single level output mode (Intensified image dissector, digital TV system).
(2) Photon counting imager without buffer memory and with parallel single level output mode (Parallel output Digicon).
(3) Photon counting imager without buffer memory and with sequential multilevel output mode (Self-scanned Digicon).

The basic image detectors quoted in the classification given above are detailed briefly in Section 3 and the typical figures for the main parameters of each detector are reported in Table 1 following the application-oriented organization scheme.

3. DESCRIPTION OF THE BASIC IMAGE DETECTORS

3.1 Multiplex image detectors

Image scanner. The time multiplexing image scanner is realized by sharing the single channel detector available in the system over the image pixels addressed by an externally controlled image scanning device.

The image scanning device may be an opto-mechanical one like a moving mirror-slit assembly, or it may be an electron beam device like in the image dissector tube where an electron multiplier section with a small acceptance area of electrically controlled position is used to accept the photoelectrons generated locally by a two dimensional photocathode (ITT Image dissector tube data sheet, 1976). It must be emphasized that the overall efficiency of an image scanner results 1/N times lower than the efficiency of the single channel detector used in the system where N is the total number of pixels contained in the image. Consequently, the image scanner is not competitive on low level, large format images if dependable multi element detectors are available on the same spectral range.

Hadamard imager. The spatial frequency multiplexing imager based on the Hadamard pseudo-binary and (0,1) normalized transformation is realized by collecting on the single channel detector used in the system the photons from the whole image area as it is seen through a series of shifts of one single cyclic type Hadamard coded mask, or through a series of basic Hadamard masks. An image of N pixels format requires a number of basic masks or shifts given by the equation $N = 2^n$, where n is the number concerned (Beauchamp 1975).

The detector outputs to the number of masks or mask shifts is a vector of measurements that constitutes the Hadamard transform of the detected image. The numerical inverse transformation yields the detected image.

The basic advantage of the Hadamard imager against the image scanner takes place when the system noise exceeds the signal noise and the full multiplex advantage is available. The efficiency of the Hadamard imager approximates in this case n/N times the efficiency of the single channel detector used.

Consequently, the Hadamard imager represents the best solution for the spectral ranges where no multi element detector is available, like the IR and far IR bands.

3.2 Analog multielement image detectors

Photographic detectors. The photographic film is a two dimensional array of randomly distributed, random size photochemical sensors with internal crosstalk. The model of the photographic detection process is a very complicated one if all the relevant system features are to be taken into account for. A simplified working model is characterized by a non-linear, thresholded density-exposure and density-wavelength behaviour with density dependent noise and spatial resolution (Dainty and Shaw 1974).

The main advantages of the photographic film are the large format, the high spatial resolution, the proven reliability, and the easy use in the UV to R spectral range. The main disadvantage consists in the non-linear detection model, that implies a complicated data processing. Further disadvantages are the low detection efficiency and the impossibility of direct calibration of the film target.

VIDICON image tube. The VIDICON is one of the first image tubes and it is a very simple and effective photoelectric imager. The device is realized by a two-dimensional, analog integrating target in the form of a photoconductive layer or of an array of silicon photodiodes that is pre-charged before the exposure and read out after the exposure by means of an electron beam of electrically controlled position. The re-charge pulses, sensed sequentially by the electron beam when each image pixel is addressed on the target, result proportional to the number of photons detected by the pixel sensed (THOMSON-CSF Image tubes catalog, 1975). The spectral range of the VIDICON and its good general performance maintain this tube still competitive for R and near IR astronomical applications.

RETICON solid state array. The RETICON all solid-state array of photodiodes and integrated read-out circuity is constituted by a one-or two-dimensional array of silicon photodiodes connected to a common video output line by individual MOSFET switches activated by externally clocking an integrated shift register for the sequential read-out of the charges analog integrated on the photodiodes capacity during the exposure time (RETICON catalog, 1975).

The RETICON technology allows to obtain a broad spectral range from the UV to the IR with a peak quantum efficiency of 80%, but at the expense of a high video output capacitance and read-out noise. This does not allow the realization of large format arrays and the use of the arrays on long integration times. The maximum number of pixels is of some 2000 for linear and 50x50 for square array formats. The maximum integration time is about one minute for cooled devices. The basic advantage of the RETICON array and of similar all solid state devices consists in the ultimate spatial stability, a feature of fundamental importance for high resolution astronomical applications on medium level signals (Vogt *et al.* 1978; Livingstone 1976).

CCD solid state array. The CCD all solid state detector is constituted by an array of photosensitive, analog integrating CCD cells organized in a shift register structure. The charges integrated in each CCD cell during the exposure time are read out by shifting them out through external clocking of the shift register. The CCD technology is compatible to both one- and two-dimensional large format arrays (RCA SID 5201 Data sheet, 1977). Up to 2500 linear and 512 x 384 rectangular array formats are commercially available and 800 x 800 pixels square format arrays are available on a tentative basis.

The main advantage of the CCD technology is the very low noise that allows the use of long integration times, up to one hour and more for cooled devices. The main disadvantage is the reduced spectral range which corresponds to operating the device in the B to IR band with the impossibility of UV operations. The CCD imagers are very suitable for astronomical applications due to their high positional stability and long integration time capability in a large format device.

CID solid state array. The CID all solid state detector is a unique modification of the CCD technology that allows to implement a non-destructive read-out of the array during the integration time. This feature is very important to verify the state of the measurement without interrupting it (TN 2000 CID Data sheet, 1977).

The CID array shows the highest intensity dynamic range available in the analog detectors field. The spatial resolution is comparable but lower than the CCD figure as well as the realizable formats that actually do not exceed a 244 x 188 pixels figure. The spectral range is the same as in the CCD array.

The non-destructive read-out feature, the high dynamic range, and the low noise make the CID device competitive for some astronomical applications and very promising for the near future (Aikens *et al.* 1976).

Electronographic detectors. The photographic emulsion operated in electron input mode shows the basic advantage with respect to the photon input mode of a linear, not thresholded density-exposure behaviour (Dainty and Shaw 1974).

An electronographic detector is realized by an image intensifier of the linear accelerator type with an analog integrating photographic film of the nuclear type as the output section of the system.

The photographic film can be mounted directly within the intensifier structure by means of a special vacuum-tight mounting, or it can be pressed on a very thin mica layer that constitutes the vacuum barrier of the system while being transparent to the output electrons. Both the solutions have been used successfully in astronomical applications (Mc Gee 1973; Mc Mullan 1976). The performance of the electronographic detector is limited by the performance of the input image intensifier section (See Section 3.4). For high gains of the image intensifier the efficiency approximates the efficiency of a photoelectric detector.

The main advantage of the electronographic approach is the realizability of large format, high efficiency detectors of high spatial resolution and good linearity by means of proven technologies.

The main disadvantage follows from the use of a photographic emulsion in the system. This yields serious problems in the calibration and in the data processing (Pilkington 1976).

SEC and EBS image tubes. The target of a VIDICON-like image tube can be operated in electron input mode with the basic advantage with respect to the photon input mode of a target gain greater than one and as high as 10^3 charges/electron for available devices. The SEC and EBS tubes are realized by coupling a semiconductive layer target and a silicon photodiodes-array target, respectively, to an image intensifier coupled to the target directly at the electron image level.

The operation of a SEC or of an EBS tube is similar to the operation of a VIDICON tube. However, the image intensifier section is responsible for an image distortion that must be calibrated against.

The basic advantage of the SEC and EBS tubes against the VIDICON tube consists in the higher responsivity associated with the target gain and in the subsequent improvement of the read-out signal-to-noise ratio (THOMSON-CSF Image tubes catalog, 1975, WESTINGHOUSE Imaging devices catalog, 1976).

The SEC tube has been applied successfully in many terrestrial and space borne astronomical experiments (Lowrance *et al.* 1976).

3.3 Digital multielement image detectors.

Intensified image dissector system. This system constitutes a two dimensional photon counter realized by optically coupling an image intensifier with phosphor output to an image dissector. At sufficiently high gain of the image intensifier each detected photon generates a flash on the output phosphor that can be detected by the image dissector operated at a frame time comparable to the phosphor decay time.

The pulses output from the image dissector are stored for each read-out to the image dissector address by means of a computer that controls the image dissector (Rybsky *et al.* 1976).

The main disadvantage of this system, which is one of the first photon counting imagers to have been realized and used in astronomical applications, consists in the small format allowed by available image dissector tubes.

Digital TV system. The system is a two-dimensional photon counter realized by coupling a high gain image intensifier with phosphor output to a VIDICON-like TV camera tube for the detection of the flashes generated by the detected photons on the phosphor screen. The basic advantage with respect to the intensified image dissector system consists in the use of the parallel output detector. This allows to integrate the phosphor output scenary on the TV camera target during the read-out frame time and to identify and reject spurious events as well as to improve the system spatial resolution by means of a real time morphological analysis of the detected events (Boksenberg 1976).

The main advantages of the approach are the large format, the high spatial resolution and the low noise.

The main disadvantage consists in the high dead time which is equal to the frame read-out time. This limits the applications of the system to low level signals only.

Self-scanned DIGICON. The self-scanned DIGICON is a one-or two-dimensional photon counter realized by electron-coupling a solid state array operated in EBS mode to an image intensifier (Choisser 1976; Tull 1976; Walker 1976).

The basic advantage of the approach consists in the target gain greater than one that is allowed by operating the array in the EBS mode and in the subsequent improvement of the read-out signal-to-noise ratio. For sufficiently high gain of the image intensifier the read-out noise becomes lower than the single photon charge packet and the system can be used in photon noise limited mode.

As the system efficiency is set by the exposure time to read-out time ratio it is necessary to integrate many events in each pixel during one single exposure. The number of events per pixel is determined after the read-out by normalizing the charge stored in each pixel to the single event charge by means of a multi-level discriminator realized by a long word A/D converter.

The number of photoevents from each read-out are stored by a computer that controls also the clocking of the solid state array used and the operation of the A/D converter.

The self-scanned DIGICONS are very effective devices which show photon noise limited performance, easy use and image intensifier limited performance in regard of image format and spatial resolution up to the limits set by the array used in the output section.

Parallel output DIGICON. The parallel output DIGICON is a one- or two-dimensional photon counter realized by electron-coupling an image intensifier to a special parallel output anode section that allows parallel pulse read-out and processing.

Several types of output sections and of processing electronics have been successfully realized and tested, including the metallic wire grid, the resistive anode, and the four quadrant anode (Kellogg 1976; Lampton 1976; Timothy 1976). The basic advantage of the parallel output DIGICON consists in an electronics limited dead time that can be as short as one microsecond with current devices. This allows to operate the system also on medium level images and not to limit the applications to faintest image cases.

The main disadvantage is the cost and complexity of the parallel processing electronics required to operate this system.

3.4 Image Intensifiers

All the multielement image detectors with photon operated targets can be coupled to an image intensifier with phosphor output to improve the overall performance on lower level images. All the image detectors with electron operated targets are actually realized around an image intensifier and are limited by the image intensifier performance.

The performance of the available image intensifiers looks to limit the ultimate performance of the image detection systems on faintest sources. The basic limitations of first-generation image intensifiers consisted in the relatively low spatial resolution and high image distortion and sensitivity to the environment. The linear accelerator type image intensifier allowed to obtain high spatial resolutions and low image distortion while being in any case sensitive to external environment. Moreover, this type of intensifier shows high voltage supply problems. (EMI Image intensifier systems catalog, 1975; THOMSON-CSF Image tubes catalog, 1975).

The realization of microchannel array plates (MCP), two-dimensional electron multipliers with pixel size comparable to that of current multielement photoelectric detectors, allowed the realization of compact, high gain image intensifiers that can be used at relatively low high voltage supply (2.5 KV) (THOMSON-CSF Images tubes catalog, 1975).

Finally, the realization of proximity focused image intensifiers allowed the realization of very compact, practically image distortion-free and environment-insensitive image intensifiers of very high gain and good spatial resolution (Timothy 1976).

The MCP proximity focused image intensifier should represent one major improvement in the performance of future image detection systems.

4. CALIBRATION PROCEDURES

The calibration of an image detector is a fundamental component of the detection process as defined in Section 1.

The calibration is concerned with all the geometrical, photometric, spectral, temporal, and operational parameters quoted in the application-oriented classification scheme given in Section 2.

Any image detector requires a calibration procedure consistent with the characteristics of the detector considered. Consequently, no consolidated standard calibration procedure is currently available which can be used for all image detectors. However, any image detector must be calibrated photometrically and spectrally and any photoelectric or electronographic image detector, which includes an image intensifier, i.e. an electron image transport section, must be calibrated against the image distortion within the system.

The more or less standard procedures used for the calibrations quoted above are described briefly in the following for the photographic and photoelectric detectors,respectively.

Calibration of the photographic detectors. It must be emphasized that the calibration of a photographic film is made unavoidably by the sample technique. This follows from the impossibility of calibrating just that film portion which is used successively for the execution of the measurement. The final accuracy of the results is then depend-

DETECTOR	IMAGE FORMAT	PIXEL FORMAT	DYNAMIC RANGE	QUANTUM EFFIC.	SPECTRAL RANGE	NOISE
IMAGE SCANNER	EFFICIENCY (1/N) TIMES DETECTOR EFFICIENCY ON N PIXELS IMAGES					
HADAMARD IMAGER	EFFICIENCY ($\log_2 N/N$) TIMES DETECTOR EFFICIENCY ON N PIXELS IMAGES^2					
PHOTOGRAPHIC FILM	30000x 30000	10x10	3.10^2	3	0.25-0.65	0.02x (1+D)
VIDICON TUBE	1000x 1000	35x35	10^2	20 80(Si)	0.3-1.05	3.10^3
RETICON ARRAY	1024x1 50x50	25x400 25x25	10^3	80	0.3-1.05	10^3
CCD ARRAY	2500x1 512x384	20x20	10^3	60	0.45-1.05	10^2
CID ARRAY	244x188	60x80	10^4	40	0.45-1.05	5.10^2
ELECTRONOGRAPHY	10000x 10000	10x10	10^3	20	0.25-0.9	0.02x (1+D)
SEC TUBE	2000x 2000	25x25	10^2	20	0.25-0.9	3
EBS TUBE	2000x 2000	25x25	10^2	20	0.25-0.9	3.10^4
INTENSIFIED IMAGE DISSECTOR	64x64	50x50	10^4	20	0.25-0.9	10^2
DIGITAL TV SYSTEM	700x700	50x50	10^4	20	0.25-0.9	10^2
SELF SCANNED DIGICON	1024x1 512x512	25x400 20x20	4.10^3	20	0.25-0.9	10^3
PARALLEL OUTPUT DIGICON	512x512	35x35	2.10^5	20	0.25-0.9	10^2
UNITY	PIXEL	MICRON	MAX OUT/ NOISE	%	MICRON	e^- rms/ frame

Table 1

ent on the validity of the extrapolation of the sample calibration to the film portion actually used for the measurement.

The photographic film is usually calibrated in regard to the density-exposure curve and to the spectral response, only. The spatial frequency response and the photochemical feedback effects are calibrated only in special cases. Of course, the typical figures of all the relevant parameters of the photographic film are given by the film manufacturer on a sample calibration technique basis but these figures are only indicative due to the poor repetitivity of the film processing.

The density-exposure calibration is performed by exposing the film sample to a series of calibrated exposure levels which are normally realized by means of a standard source and a series of calibrated attenuators.

The spectral responsivity is calibrated in a similar mode by previously dispersing the output from the standard source by means of a spectrograph.

The spatial frequency response and the photochemical feedback effects are usually calibrated by exposing the sample film to suitable test patterns and then performing a proper data processing.

A comprehensive description of the calibration procedures for the photographic film is given in the Image Science book (Dainty and Shaw 1974) and for electronography by Pilkington (1976).

Calibration of the photoelectric detectors. It must be emphasized that the calibration of a photoelectric detector is performed on the same target which is used successively for the execution of the measurement. The final accuracy of the results is then dependent substantially on the stability of the calibration against time.

The photoelectric detectors are usually calibrated in regard to the photometric input-output curve and to the spectral response by the same technique quoted above for the photographic film.

Also the spatial frequency response is determined in a similar mode. It must be emphasized that the photoelectric detectors are compatible to a local calibration of the detector target implemented on a pixel per pixel basis. This feature is responsible for the high quality of the photometric data from photoelectric image detectors.

A negative feature of any photoelectric detector which includes an electron image transport section, like any intensified detector, is the image distortion taking place within the transport section. This image distortion must be calibrated to allow for further data correction. The calibration of the image distortion is usually performed by exposing the detector to a test pattern of known geometry.

Some partial descriptions of the calibration procedures and of the associated data processing for some basic photoelectric image detectors are given by Fort *et al.* (1976), Cenalmor *et al.* (1976), Lamy *et al.*(1976), Gow *et al.* (1976) and Walker *et al.* (1976).

5. COMPARISON OF CURRENT STATE OF THE ART IMAGE DETECTORS. TRENDS OF DEVELOPMENT FOR FUTURE DETECTORS

A comparison of the current state-of-the-art image detectors can now be made as a scientific application-oriented exercise on the basis of the data given in Table 1.

Once emphasized that all the image detectors quoted in Table 1 have been tested successfully in astronomical applications, it is correct to point out that the photoelectric image detectors are clearly going to outperform the photographic and electronographic detectors in high-accuracy, low-level two-dimensional photometry.

The photographic detector remains the best detector available for large image format applications which require high geometrical stability, such as wide field astrometry.

Among the photoelectric detectors the SEC image tube seems to be the best large format, multielement analog integrating detector available now. The digital TV system seems to be the best large format, multi element image photon counter.

The multielement detectors seem to have outperformed the image scanners and the spatial frequency multiplexing imagers in the spectral range from the UV to the near IR up to 1 micron wavelength. However, the Hadamard transform imager seems to be unbeatable for IR and far IR application where the system noise is dominated by the detector noise. Some further considerations are needed to evaluate correctly the electronographic detectors. The electronographic detectors show photoelectric detector-like quantum efficiency and spectral response but they cannot of course outperform the efficiency of the image intensifier that constitutes their input section. The basic advantage consists in the use of a large format, multielement output section realized by the nuclear emulsion film used as the analog integrating detector. The current availability of SEC and EBS targets of format consistent with the output format of available image intensifiers is going to overcome this basic advantage of the electronographic approach, which of course still remains perfectly valid for image formats exceeding the available formats of the photoelectric targets.

Regarding the future the self-scanned Digicon-like image detectors with larger format CCD, CID or avalanche photodiodes solid-state arrays as the output section and the parallel output Digicon-like detectors with integrated pulse processors arrays at the output section appear to be the most promising devices for high accuracy, low level two-dimensional photometry.

The use of the latest generation double proximity focussed MCP image intensifiers should extend the application field of the photoelectric multielement detectors to high geometrical accuracy applications, thus probably outperforming the photographic film also in astrometrical applications.

In the IR and perhaps also in the far IR spectral range the realization of multielement solid state detectors similar to the existing arrays should provide for an efficiency higher than that allowed by the Hadamard transform imager.

The greatest problem foreseen lies in the data storage at archival level. The use of optical memories of virtually unlimited capacity should provide for suitable storage means, which have nothing different from the existing photographic film stores, while showing the advantage of storing the information directly in the form of calibrated arrays of numbers directly addressable by a computer.

As a final consideration it is necessary to emphasize that the increasing importance of space astronomy yields a concentration of efforts towards the realization of higher performance, higher efficiency image detectors that can be operated under computer control. It is reasonable to extrapolate the present situation into a future, but a relatively near one, where the photoelectric detectors will have definitely gained the best position among the optical image detection systems.

REFERENCES

Aikens R.S., Harvey J.W., Lynds C.R.: 1976, IAU Colloq. No. 40, 25, Eds. M. Duchesne, G. Lelievre, Paris-Meudon Obs.

Beauchamp K.G.: 1975, *Walsh functions and their applications*, Academic Press, London.

Boksenberg A.: 1976, IAU Colloq. No. 40, 13.

Cenalmore V.: 1976, IAU Colloq. No. 40, 16.

Choisser J.P.: 1976, IAU Colloq. No. 40, 27.

Dainty J.C., Shaw R.: 1974, *Image science*, Academic Press, London.

De Batz B., Bensammar S., Delavand J., Gay J., Journet A.: 1977, Infrared Phys. 17, 305.

EMI Image intensifier systems catalog, 1975, EMI Electronics Ltd, Hayes, Middlesex, England.

Fort B., Boksenberg A., Coleman C.: 1976, IAU Colloq. No 40, 15.

Gow C.E., Sandford II M.T., Honeycutt R.K., Jekowski J.P.: 1976, IAU Colloq. No. 40, 21.

Hansen P., Strong J.: 1972, Appl. Opt. 11, 502.

Kellogg E., Henry P., Murray S., Van Speybroeck L.: 1976, Rev.Sci. Instrum. 47, 282.

ITT Image dissector data sheet, 1976, ITT Electro optical products div., Roanoke, Virginia USA

Lamy P.L., Nguyen-Trong T., Perrin J.M., 1976, IAU Colloq. No. 40, 17.

Lampton M., 1976, IAU Colloq. No. 40, 32.

Livingston W.C.: 1976, IAU Colloq. No. 40, 22.

Lowrance J.L., Zucchino P., Williams T.B.: 1976, IAU Colloq. No. 40, 18.

Mc Gee J.D.: 1973, Vistas in Astronomy 15, 61.

Mc Mullan D., Powell J.R.: 1976, IAU Colloq. No. 40, 5.

Mc Nall J.F., Nordsieck K.H.: 1976, IAU Colloq. No. 40, 26.

Pilkington J.D.H.: 1976, IAU Colloq. No. 40, 10.

RCA SID5201 CCD imager data sheet, RCA, Lancaster, Pennsylvania, USA.

Rybsky P.M., Van Cittern G.W., Benedict G.F.: 1976, IAU Colloq. No. 40, 54.

RETICON solid state image sensors and systems catalog, 1975, Reticon Co., Mountain View, California, USA.

Timothy J.G.: 1976, IAU Colloq. No. 40, 33.

THOMSON CSF Image tubes catalog, 1975, Thomson CSF, Paris, France.

TN2000 CID Imager data sheet, General Electric Optoelectronic Systems, Syracuse, New York, USA.

Tull R.G.: 1976, IAU Colloq. No. 40, 23.

Vogt S.S., Tull R.G., Kelton P.: 1978, Appl. Opt. 17, 574.

Walker G.A.H., Buchholz V., Fahlman G.G., Glaspey J., Lane-Wright D., Mochnacki S.: 1976, IAU Colloq. No. 40, 24.

WESTINGHOUSE Imaging devices catalog, 1976, Westinghouse Electric Co., New York, USA.

THE TEACHING OF ASTRONOMY TO THE SWEDISH PEOPLE

Aage Sandqvist
Stockholm Observatory, Saltsjöbaden, Sweden

ABSTRACT

The teaching of astronomy to the Swedish people is accomplished on many different levels and has, in the last couple of years, passed through several revolutionary stages of development. At the pre-university level, astronomy often accompanies the physics curriculum, and practical experience is made available to the more promising middle and high school student through an apprentice system which enables the student to work at an observatory for several weeks. The major reform of the Swedish University system, which took place in the fall of 1977, was preceded two years earlier by a countrywide reform of the astronomy courses offered at the university undergraduate level, a reform which has opened up astronomy to other scientific and mathematical disciplines, other university faculties and the general public. A novel approach to the dissemination of astronomy to the general public has been started in 1977 through the creation of the "Information project - the universe", which has already reached hundreds of thousands of persons and will touch many more in its continuing efforts during the next few years.

ELEMENTARY AND SECONDARY SCHOOLS

The child in Sweden normally begins school when he is seven years old, often having spent a year or more in some sort of kindergarten beforehand. In his first three school years he takes a course in "regional studies". This is an integrated course dealing with the child's experience of science and society. Astronomy comes into the picture through simple presentations of the seasons of the year, the clock, the almanack and so on. In school years 4 through 6, the child takes "natural science", amongst other courses. One seventh, or 37 hours, of this integrated course deals with physics and here a slight amount of astronomy is also taught, the actual amount depending mainly upon the interest of the teacher. In school years 7 through 9, about 20 hours of astronomy is taught inside the frame of physics. Teachers

Bengt E. Westerlund (ed.), Stars and Star Systems, 235–247.

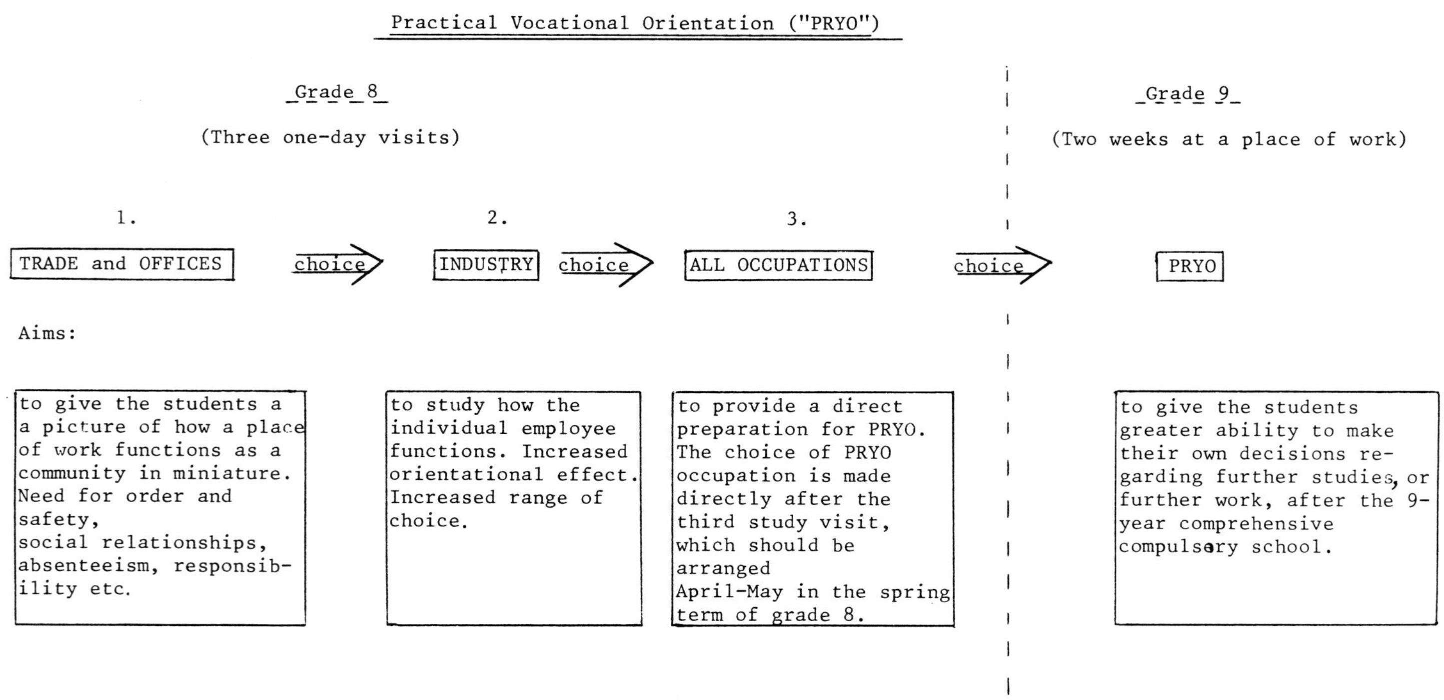

Fig. 1. Schematic presentation of the Swedish apprenticeship system (PRYO), in which the Swedish astronomical observatories participate.

are encouraged to teach natural sciences in form of an integrated science curriculum, but presently only 20 percent of the teaching takes place in the form of integrated science due to psychological resistance amongst the teachers and the threat of more work. In school years 10 through 12, which are not compulsory, astronomy is amply represented through examples of application of physics and through special courses. Both radio and television are being used to introduce astronomy into the schools and programs have proven to be popular.

APPRENTICESHIP

It is in grades 8 and 9 that the student is confronted with the apprenticeship system, known as practical vocational orientation ("PRYO"). The primary aim of the system is to give the student an orientation in outside working life at large, but it also offers a picture of the working duties of individuals in different occupations. This is accomplished by at least three compulsory study visits in grade 8 to stores and offices, industries and other fields, followed by two weeks at places of work in the autumn term of grade 9. (Fig. 1). The student is given the choice of field for his own two-week PRYO period after the third study visit, but because of the popularity of the observatories and the great demand thus placed on them, only students with high academic standings are accepted for the astronomy PRYO.

By means of the three study visits, which are spread over a year, the students are trained in gathering information from different sources, making observations during the actual visit to the firm and selecting, evaluating and comparing their impressions, and setting them in relation to their expectations. They also come into contact with three different occupational environments. The astronomical observatories, however, do not partake in the phase involving the study visits in grade 8 but make themselves available for the extended work visit in grade 9. There can be no denying that this places an extra burden on astronomers as they have to prepare meaningful work for these young students. But we consider it well worth our efforts to make the contacts with these promising youngsters, who may someday choose astronomy as their field of endeavour, or at least be favourably disposed towards it.

UNIVERSITY

In the fall of 1977 a major reform of the Swedish University system took place. It is hoped that this reform will effect an equalization of the possibilities that members of different social classes may have of attending universities and, furthermore, strengthen ties between society and higher learning. To accomplish this, the universities were reorganized administratively and at the same time

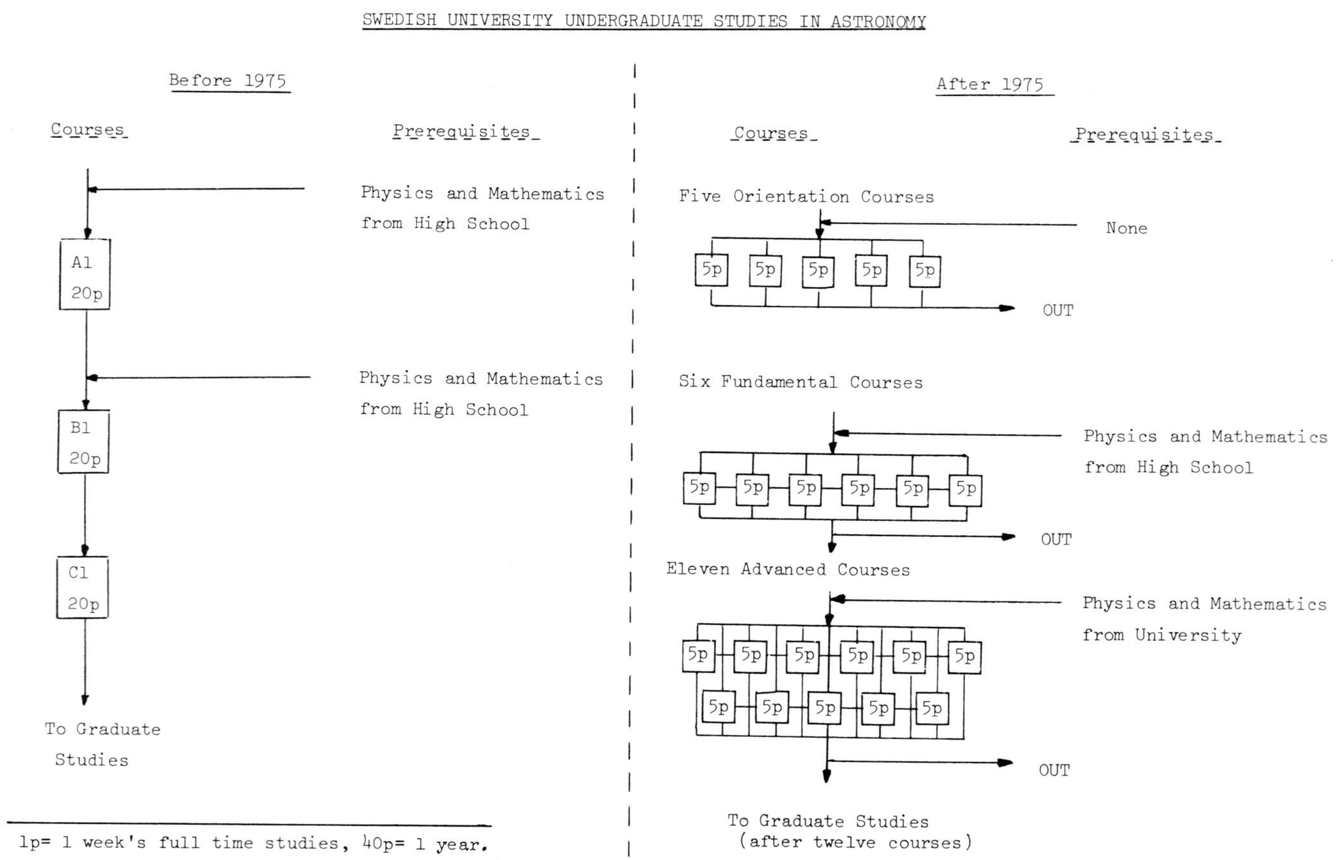

Fig. 2. Schematic presentation of Swedish university undergraduate courses in astronomy, before and after 1975.

opened up to everyone who is at least 25 years old and has four or more years working experience (includes housewives and househusbands). A major emphasis is placed on the possibility of returning to university several times during a lifetime for the updating of knowledge or the continuation of a course of study. Also, job- or career-oriented curricula receive high ratings of importance.

With a watchful eye towards these developments, the institutes of astronomy in Sweden enacted, in the fall of 1975, their own country-wide reform of the astronomy courses offered at the university undergraduate level. (Fig. 2). The previous system emphasized a course of study which lead mainly to graduate studies in astronomy, but the new system (which can be used to reconstruct the old system) opens up astronomy in packageformat to (i) other disciplines such as physics, chemistry and mathematics (ii) other faculties such as arts and the social sciences and (iii) the general public.

The system consists of 22 courses each of which is worth 5 points (one point corresponds to one week of full-time studies and one year's full-time study normally results in 40 points). A course is most often given at halfspeed, lasts 10 weeks and can be read in parallel with another course. The student may acquire anywhere from 5 to 60 points of astronomy depending upon his interests and completed prerequisites. To begin graduate studies in astronomy an undergraduate degree of 120 points, containing 60 points of astronomy (which may not include courses 19-23), must first be obtained. The courses and their prerequisites are given in the following table:

No.	Name	Prerequisites in Mathematics and Physics
1.	Outline of astronomy	secondary school
2.	Introduction to astronomy	" "
3.	Instrumental technique	" "
4.	Astrophysics	" "
5.	Galaxies	" "
6.	Spherical and practical astronomy	" "
7.	Advanced instrumental technique	university
8.	Celestial mechanics	"
9.	Physics of the planetary system	"
10.	Stellar atmospheres	"
11.	Stellar structure and evolution	"
12.	The Milky Way system	"
13.	Physics of the interstellar medium	"
14.	Stellar dynamics	"
15.	Cosmology	"
16.	Special topics 1	variable
17.	Special topics 2	"
18.	(Administrative jugglebox)	"
19.	Orientation in astronomy	no prerequisites at all

20.	Historical development of concepts of the universe	no prerequisites at all
21.	The structure of the universe	"
22.	Special topics 3	"
23.	Special topics 4	"

Course 1 is intended predominantly for school teachers of physics and astronomy, course 16, 17, 22 and 23 have flexible content and may deal with topics of special or current interest e.g. the sun, origin of the elements, a research project, or life in the universe. The prerequisites in mathematics and physics at the university level are generally 20 and 40 points, respectively, but the astronomy courses may usually be taken without any other course in astronomy being a prerequisite. (This excepts the astronomy student who wishes to take 60 points and go on to graduate studies; he is advised to take the courses in a loose order.) The aspired result of this scheme is e.g. that an advanced physics student may take course 11, a chemistry student course 13, a mathematics student course 8, a history student course 20, etc.

The new system has been successful - the number of undergraduate students in astronomy has increased by a factor of ten. This is predominantly due to the orientation courses which have no prerequisites, but even at the higher levels there has been a significant increase in the number of students. The great majority, by far, no longer studies astronomy with the intention of going on to graduate studies but is motivated by an intense interest in the subject.

Graduate work in astronomy may be carried out by the student who has earned an undergraduate degree which contains 60 points in astronomy. An additional 70 points of course work then has to be amassed, and on the successful completion of a doctoral dissertation, representing original research work, the student is awarded a "fil.dr" (Ph.D.) degree. In theory, this is supposed to take four years, but experience has shown that five or six years is a more reasonable estimate.

INFORMATION PROJECT "THE UNIVERSE"

The information project "The Universe" is an attempt to eliminate the severe gap that has existed in Sweden between the general public's horrific appetite for astronomy and this subject's previously very limited accessibility. The recent years' spectacular events in space technology has whetted the public's appetite and caused a drastic rise of interest in the universe. However, the massmedia covered the news about Moon- and Mars-trips mainly from a technical viewpoint and rarely followed up the scientific results. This is the background for the creation of the information project in the fall of 1977.

The information project "The Universe" is a co-operative project which includes the following participants:

Swedish Natural Science Research Council/Editorial Service (co-ordinator)
The popular scientific magazine "Forskning och Framsteg"
Swedish Museum of Natural History
Royal Swedish Academy of Sciences
Swedish Travelling Exhibition
Museum of Technology
Stockholm Observatory
Onsala Space Observatory
Swedish Space Corporation
Swedish Board for Space Activities
Swedish National Committee of Astronomy
Swedish Astronomical Society
Amateur Astronomers in Sweden
The daily newspaper "Dagens Nyheter"
Swedish Collective Book Promotion
Swedish Broadcasting Corporation
Organisations for general and adult education
The universities in Sweden

A number of organisations, firms, etc. are backing the project with resources such as money, skill, working-power, different material and exhibits. Contacts have also been established with firms such as Rockwell, SAAB Aerospace Div, Volvo, Zeiss, Hasselblad. The principle for financing the project is that each participant carries his own costs and only the coordinating administration and the general marketing should be funded from special grants. The project started in August 1977 with an introductory phase and continued in a conspicuous phase, called the Space Week, in the middle of October. This was followed by the long-range phase which is going on now and will last for about four or five years.

The introductory phase was meant to prepare the massmedia and the public for the Space Week. It involved, for example, massive advertisements in newspapers and on posters, as well as a most successful drawing competition in Dagens Nyheter (Scandinavia's largest newspaper with a daily circulation of 450 000 copies). Children between 5 and 15 years old were invited to draw pictures of their impressions of "Life in the Universe" - and 50 000 responded. The winning entry was enlarged into a large three-dimensionel model (Fig. 3) which served as a master of ceremonies in the later phases of the information project.

The Space Week was an eight day PR-event which also introduced the long-range phase. During this week the public had the opportunity to meet informally with both amateur and professional astronomers, educationalists, rocket enthusiasts, science fiction fans, representatives of firms dealing with telescopes, cameras, books. The Space Week took place at the Wenner-Gren Centre which contains Stockholm's

Figure 3. Professor Ingvar Lindqvist (Secretary General of the Swedish Natural Science Research Council) speaking to "Little Green Man" during opening ceremonies of the information project, The Universe.

tallest building (25 floors). Various exhibits such as satellites and rockets were displayed on the main floor; here is also an auditorium where films about astronomy and space research were shown. The 23rd and 24th floors of the building had been clad in a futuristic decor and here the interaction between people took place. Amateur astronomers had set up their telescopes on the roof of the building, enabling the public to study Stockholm's cloudy skies. A number of project items went on sale at this event and are being sold in connection with the later phases also. These include an umbrella with the constellations pictured on the inside, "The Book about Space" - a booklet with easily read and comprehended text as well as many colourful pictures, and various large posters (Fig. 4). The public flocked to the Space Week in large numbers - a total of 25 000 persons, which was three times more than expected.

The long-range phase started at the end of the Space Week with one of the project's biggest parts: an exhibition at the Swedish Museum of Natural History called "We live in Cosmos".

Fig. 4. The star umbrella seen from "above". With its yellow constellations printed on a blue background, it can be used as an effective star map or a conventional umbrella, depending upon weather conditions. A small, weak flashlight is attached to the handle of the umbrella.

Fig. 5. Scenes from the exhibition "We live in Cosmos" (see pages 244-246).

a) The famous Lund representation of the Milky Way in its "natural" size. A luminous three dimensional model of the Galaxy is rotating in the foreground.

b) A three dimensional model, fluorescing from illumination by ultra-violet light, illustrating the black hole model of the X-ray source, Cygnus X-1.

c) Section on meteorites. Exhibited samples include one weighing 102 kg which lies under a glass cover in the table top.

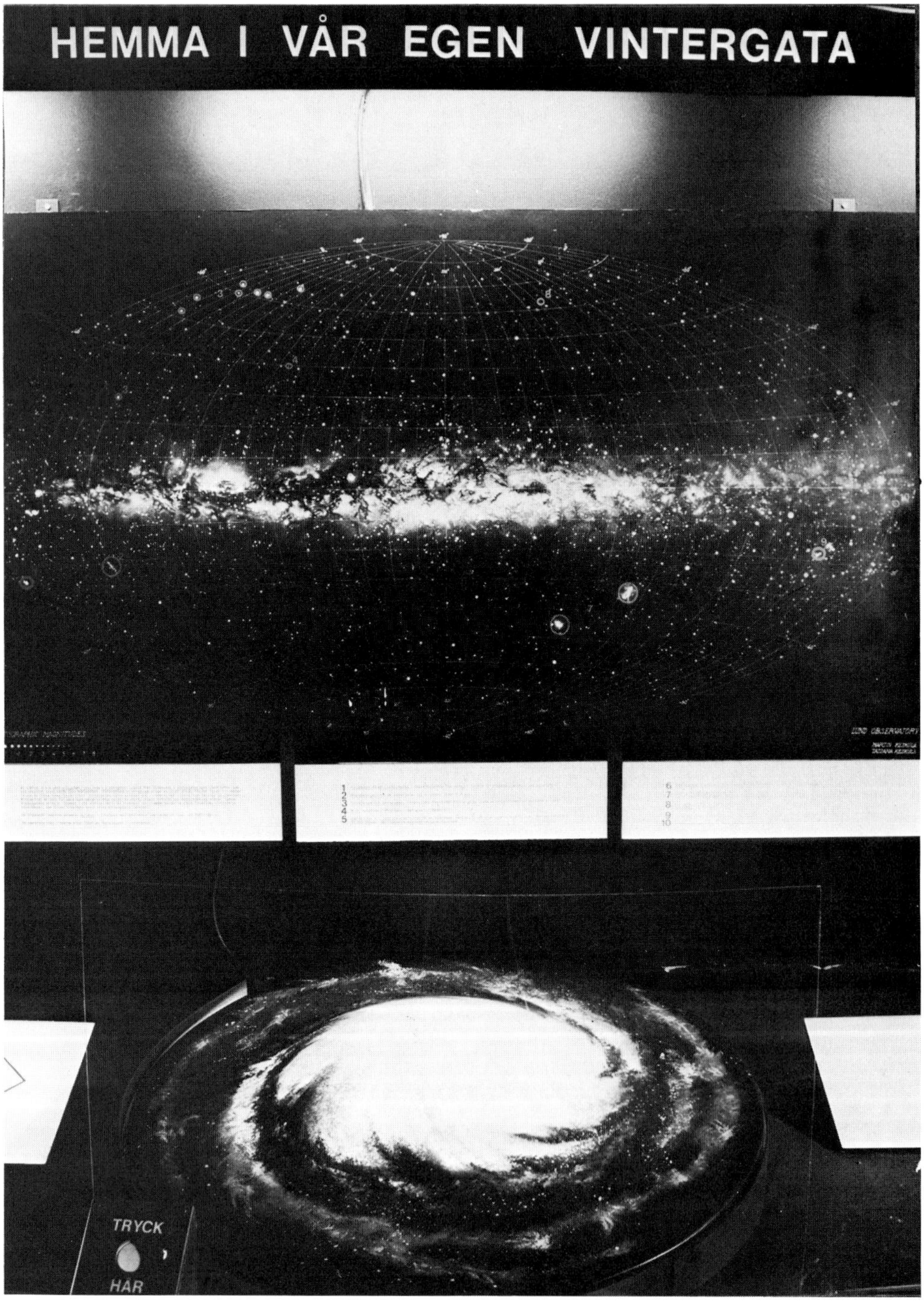

Figure 5 a

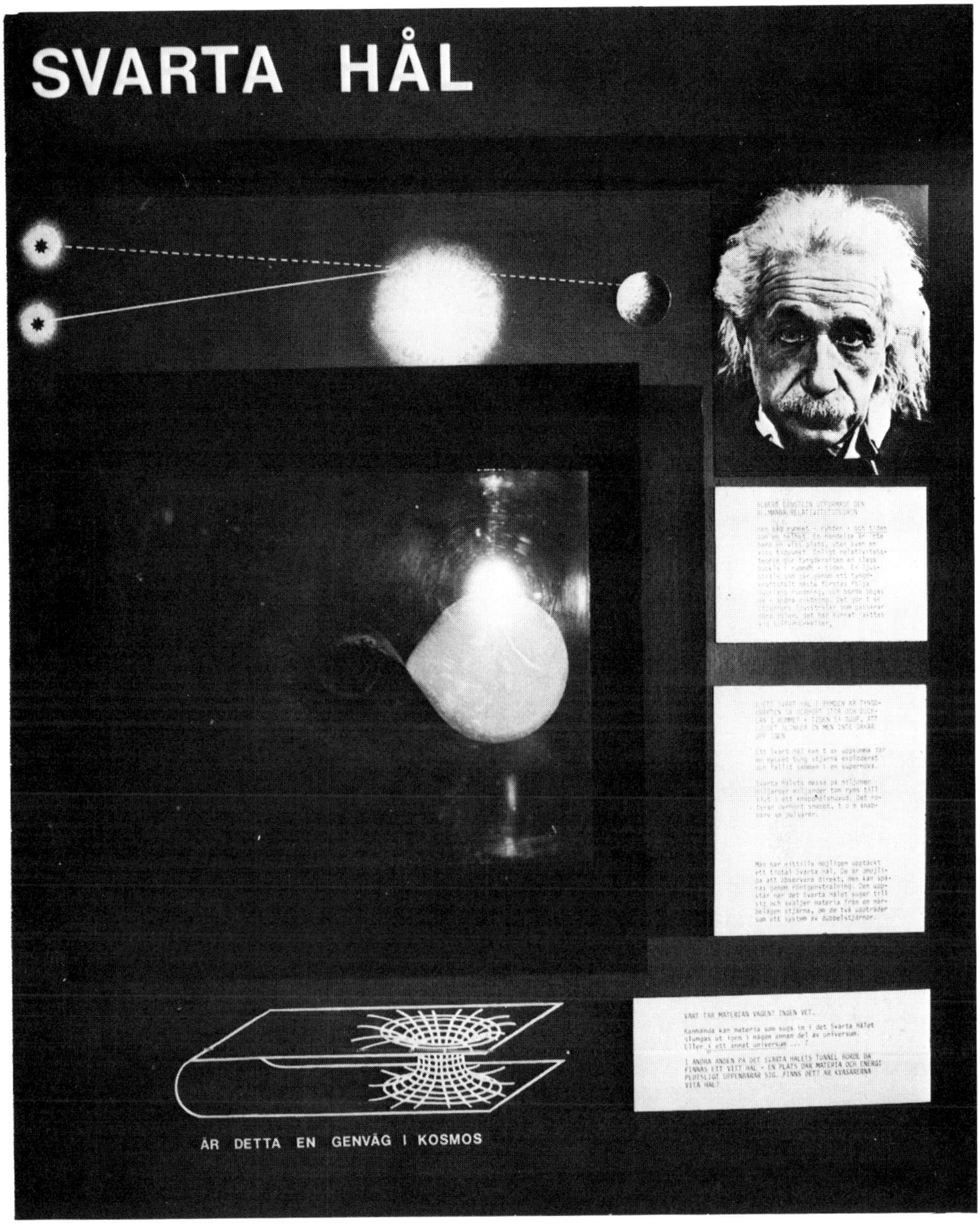

Figure 5 b

METEORITER – STENAR FRÅN RYMDEN
SILJANSRINGEN
EUROPAS STÖRS
KÄNDA
METEORITNEDSLAG
INTRÄFFADE FÖR
NÄRA 400 MILJONER
ÅR SEDAN

Figure 5

The exhibition begins with a historical development of concepts of the universe up to our present viewpoint. It then takes the visitor from the building blocks of the universe, the galaxies, to our Milky Way and subsequently to the stars. The solar system is treated in detail with the help of the spectacular space technology photographs. After the Earth has had its proper place in the universe described, the question of the creation of life on Earth and the possibility of life elsewhere is considered, together with possible communication techniques. Audio-visual aids and three-dimensional movable models are amply used to illustrate space travel, radio noise from pulsars, black holes, the structure of the Milky Way etc. A minuscule piece of the Moon is exhibited together with an authentic Apollo space suit, which is on loan from the Smithsonian Institute, USA. Scenes from the exhibition are shown in Fig. 5.

This exhibition, "We live in Cosmos", will stay in Stockholm for one and a half years and will then, sometime in 1979, be converted into a travelling exhibition which will move around in Sweden for several years. Each time the main exhibition comes to a new location, smaller poster exhibitions will circulate in the community like satellites around a planet. The local amateur astronomers will be activated and organize special programs and activities. In this way it is hoped that the Swedish people, even in remote areas, will have access to astronomical knowledge and will become appreciative of the universe in which we live.

SUBJECT INDEX

LIST OF INDIVIDUAL OBJECTS

GALAXIES

CLUSTERS OF GALAXIES

PULSARS

QUASARS

RADIO SOURCES

X-RAY CLUSTERS

X-RAY STARS

ASTROPHYSICS AND SPACE SCIENCE LIBRARY

Edited by

J. E. Blamont, R. L. F. Boyd, L. Goldberg, C. de Jager, Z. Kopal, G. H. Ludwig, R. Lüst, B. M. McCormac, H. E. Newell, L. I. Sedov, Z. Švestka, and W. de Graaff

1. C. de Jager (ed.), *The Solar Spectrum, Proceedings of the Symposium held at the University of Utrecht, 26–31 August, 1963.* 1965, XIV + 417 pp.
2. J. Ortner and H. Maseland (eds.), *Introduction to Solar Terrestrial Relations, Proceedings of the Summer School in Space Physics held in Alpbach, Austria, July 15–August 10, 1963 and Organized by the European Preparatory Commission for Space Research.* 1965, IX + 506 pp.
3. C. C. Chang and S. S. Huang (eds.), *Proceedings of the Plasma Space Science Symposium, held at the Catholic University of America, Washington, D.C., June 11–14, 1963.* 1965, IX + 377 pp.
4. Zdeněk Kopal, *An Introduction to the Study of the Moon.* 1966, XII + 464 pp.
5. B. M. McCormac (ed.), *Radiation Trapped in the Earth's Magnetic Field. Proceedings of the Advanced Study Institute, held at the Chr. Michelsen Institute, Bergen, Norway, August 16–September 3, 1965.* 1966, XII + 901 pp.
6. A. B. Underhill, *The Early Type Stars.* 1966, XII + 282 pp.
7. Jean Kovalevsky, *Introduction to Celestial Mechanics.* 1967, VIII + 427 pp.
8. Zdeněk Kopal and Constantine L. Goudas (eds.), *Measure of the Moon. Proceedings of the 2nd International Conference on Selenodesy and Lunar Topography, held in the University of Manchester, England, May 30–June 4, 1966.* 1967, XVIII + 479 pp.
9. J. G. Emming (ed.), *Electromagnetic Radiation in Space. Proceedings of the 3rd ESRO Summer School in Space Physics, held in Alpbach, Austria, from 19 July to 13 August, 1965.* 1968, VIII + 307 pp.
10. R. L. Carovillano, John F. McClay, and Henry R. Radoski (eds.), *Physics of the Magnetosphere, Based upon the Proceedings of the Conference held at Boston College, June 19–28, 1967.* 1968, X + 686 pp.
11. Syun-Ichi Akasofu, *Polar and Magnetospheric Substorms.* 1968, XVIII + 280 pp.
12. Peter M. Millman (ed.), *Meteorite Research. Proceedings of a Symposium on Meteorite Research, held in Vienna, Austria, 7–13 August, 1968.* 1969, XV + 941 pp.
13. Margherita Hack (ed.), *Mass Loss from Stars. Proceedings of the 2nd Trieste Colloquium on Astrophysics, 12–17 September, 1968.* 1969, XII + 345 pp.
14. N. D'Angelo (ed.), *Low-Frequency Waves and Irregularities in the Ionosphere. Proceedings of the 2nd ESRIN-ESLAB Symposium, held in Frascati, Italy, 23–27 September, 1968.* 1969, VII + 218 pp.
15. G. A. Partel (ed.), *Space Engineering. Proceedings of the 2nd International Conference on Space Engineering, held at the Fondazione Giorgio Cini, Isola di San Giorgio, Venice, Italy, May 7–10, 1969.* 1970, XI + 728 pp.
16. S. Fred Singer (ed.), *Manned Laboratories in Space. Second International Orbital Laboratory Symposium.* 1969, XIII + 133 pp.
17. B. M. McCormac (ed.), *Particles and Fields in the Magnetosphere. Symposium Organized by the Summer Advanced Study Institute, held at the University of California, Santa Barbara, Calif., August 4–15, 1969.* 1970, XI + 450 pp.
18. Jean-Claude Pecker, *Experimental Astronomy.* 1970, X + 105 pp.
19. V. Manno and D. E. Page (eds.), *Intercorrelated Satellite Observations related to Solar Events. Proceedings of the 3rd ESLAB/ESRIN Symposium held in Noordwijk, The Netherlands, September 16–19, 1969.* 1970, XVI + 627 pp.
20. L. Mansinha, D. E. Smylie, and A. E. Beck, *Earthquake Displacement Fields and the Rotation of the Earth, A NATO Advanced Study Institute Conference Organized by the Department of Geophysics, University of Western Ontario, London, Canada, June 22–28, 1969.* 1970, XI + 308 pp.
21. Jean-Claude Pecker, *Space Observatories.* 1970, XI + 120 pp.
22. L. N. Mavridis (ed.), *Structure and Evolution of the Galaxy. Proceedings of the NATO Advanced Study Institute, held in Athens, September 8–19, 1969.* 1971, VII + 312 pp.
23. A. Muller (ed.), *The Magellanic Clouds. A European Southern Observatory Presentation: Principal Prospects, Current Observational and Theoretical Approaches, and Prospects for Future Research, Based on the Symposium on the Magellanic Clouds, held in Santiago de Chile, March 1969, on the Occasion of the Dedication of the European Southern Observatory.* 1971, XII + 189 pp.

24. B. M. McCormac (ed.), *The Radiating Atmosphere. Proceedings of a Symposium Organized by the Summer Advanced Study Institute, held at Queen's University, Kingston, Ontario, August 3–14, 1970.* 1971, XI + 455 pp.
25. G. Fiocco (ed.), *Mesospheric Models and Related Experiments. Proceedings of the 4th ESRIN-ESLAB Symposium, held at Frascati, Italy, July 6–10, 1970.* 1971, VIII + 298 pp.
26. I. Atanasijević, *Selected Exercises in Galactic Astronomy.* 1971, XII + 144 pp.
27. C. J. Macris (ed.), *Physics of the Solar Corona. Proceedings of the NATO Advanced Study Institute on Physics of the Solar Corona, held at Cavouri-Vouliagmeni, Athens, Greece, 6–17 September 1970.* 1971, XII + 345 pp.
28. F. Delobeau, *The Environment of the Earth.* 1971, IX + 113 pp.
29. E. R. Dyer (general ed.), *Solar-Terrestrial Physics/1970. Proceedings of the International Symposium on Solar-Terrestrial Physics, held in Leningrad, U.S.S.R., 12–19 May 1970.* 1972, VIII + 938 pp.
30. V. Manno and J. Ring (eds.), *Infrared Detection Techniques for Space Research. Proceedings of the 5th ESLAB-ESRIN Symposium, held in Noordwijk, The Netherlands, June 8–11, 1971.* 1972, XII + 344 pp.
31. M. Lecar (ed.), *Gravitational N-Body Problem. Proceedings of IAU Colloquium No. 10, held in Cambridge, England, August 12–15, 1970.* 1972, XI + 441 pp.
32. B. M. McCormac (ed.), *Earth's Magnetospheric Processes. Proceedings of a Symposium Organized by the Summer Advanced Study Institute and Ninth ESRO Summer School, held in Cortina, Italy, August 30–September 10, 1971.* 1972, VIII + 417 pp.
33. Antonin Rükl, *Maps of Lunar Hemispheres.* 1972, V + 24 pp.
34. V. Kourganoff, *Introduction to the Physics of Stellar Interiors.* 1973, XI + 115 pp.
35. B. M. McCormac (ed.), *Physics and Chemistry of Upper Atmospheres. Proceedings of a Symposium Organized by the Summer Advanced Study Institute, held at the University of Orléans, France, July 31–August 11, 1972.* 1973, VIII + 389 pp.
36. J. D. Fernie (ed.), *Variable Stars in Globular Clusters and in Related Systems. Proceedings of the IAU Colloquium No. 21, held at the University of Toronto, Toronto, Canada, August 29–31, 1972.* 1973, IX + 234 pp.
37. R. J. L. Grard (ed.), *Photon and Particle Interaction with Surfaces in Space. Proceedings of the 6th ESLAB Symposium, held at Noordwijk, The Netherlands, 26–29 September, 1972.* 1973, XV + 577 pp.
38. Werner Israel (ed.), *Relativity, Astrophysics and Cosmology. Proceedings of the Summer School, held 14–26 August, 1972, at the BANFF Centre, BANFF, Alberta, Canada.* 1973, IX + 323 pp.
39. B. D. Tapley and V. Szebehely (eds.), *Recent Advances in Dynamical Astronomy. Proceedings of the NATO Advanced Study Institute in Dynamical Astronomy, held in Cortina d'Ampezzo, Italy, August 9–12, 1972.* 1973, XIII + 468 pp.
40. A. G. W. Cameron (ed.), *Cosmochemistry. Proceedings of the Symposium on Cosmochemistry, held at the Smithsonian Astrophysical Observatory, Cambridge, Mass., August 14–16, 1972.* 1973, X + 173 pp.
41. M. Golay, *Introduction to Astronomical Photometry.* 1974, IX + 364 pp.
42. D. E. Page (ed.), *Correlated Interplanetary and Magnetospheric Observations. Proceedings of the 7th ESLAB Symposium, held at Saulgau, W. Germany, 22–25 May, 1973.* 1974, XIV + 662 pp.
43. Riccardo Giacconi and Herbert Gursky (eds.), *X-Ray Astronomy.* 1974, X + 450 pp.
44. B. M. McCormac (ed.), *Magnetospheric Physics. Proceedings of the Advanced Summer Institute, held in Sheffield, U.K., August 1973.* 1974, VII + 399 pp.
45. C. B. Cosmovici (ed.), *Supernovae and Supernova Remnants. Proceedings of the International Conference on Supernovae, held in Lecce, Italy, May 7–11, 1973.* 1974, XVII + 387 pp.
46. A. P. Mitra, *Ionospheric Effects of Solar Flares.* 1974, XI + 294 pp.
47. S.-I. Akasofu, *Physics of Magnetospheric Substorms.* 1977, XVIII + 599 pp.
48. H. Gursky and R. Ruffini (eds.), *Neutron Stars, Black Holes and Binary X-Ray Sources.* 1975, XII + 441 pp.
49. Z. Švestka and P. Simon (eds.), *Catalog of Solar Particle Events 1955–1969. Prepared under the Auspices of Working Group 2 of the Inter-Union Commission on Solar-Terrestrial Physics.* 1975, IX + 428 pp.
50. Zdeněk Kopal and Robert W. Carder, *Mapping of the Moon.* 1974, VIII + 237 pp.
51. B. M. McCormac (ed.), *Atmospheres of Earth and the Planets. Proceedings of the Summer Advanced Study Institute, held at the University of Liège, Belgium, July 29–August 8, 1974.* 1975, VII + 454 pp.
52. V. Formisano (ed.), *The Magnetospheres of the Earth and Jupiter. Proceedings of the Neil Brice Memorial Symposium, held in Frascati, May 28–June 1, 1974.* 1975, XI + 485 pp.

53. R. Grant Athay, *The Solar Chromosphere and Corona: Quiet Sun.* 1976, XI + 504 pp.
54. C. de Jager and H. Nieuwenhuijzen (eds.), *Image Processing Techniques in Astronomy. Proceedings of a Conference, held in Utrecht on March 25–27, 1975*, XI + 418 pp.
55. N. C. Wickramasinghe and D. J. Morgan (eds.), *Solid State Astrophysics. Proceedings of a Symposium, held at the University College, Cardiff, Wales, 9–12 July 1974.* 1976, XII + 314 pp.
56. John Meaburn, *Detection and Spectrometry of Faint Light.* 1976, IX + 270 pp.
57. K. Knott and B. Battrick (eds.), *The Scientific Satellite Programme during the International Magnetospheric Study. Proceedings of the 10th ESLAB Symposium, held at Vienna, Austria, 10–13 June 1975.* 1976, XV + 464 pp.
58. B. M. McCormac (ed.), *Magnetospheric Particles and Fields. Proceedings of the Summer Advanced Study School, held in Graz, Austria, August 4–15, 1975.* 1976, VII + 331 pp.
59. B. S. P. Shen and M. Merker (eds.), *Spallation Nuclear Reactions and Their Applications.* 1976, VIII + 235 pp.
60. Walter S. Fitch (ed.), *Multiple Periodic Variable Stars. Proceedings of the International Astronomical Union Colloquium No. 29, Held at Budapest, Hungary, 1–5 September 1975.* 1976, XIV + 348 pp.
61. J. J. Burger, A. Pedersen, and B. Battrick (eds.), *Atmospheric Physics from Spacelab. Proceedings of the 11th ESLAB Symposium, Organized by the Space Science Department of the European Space Agency, held at Frascati, Italy, 11–14 May 1976.* 1976, XX + 409 pp.
62. J. Derral Mulholland (ed.), *Scientific Applications of Lunar Laser Ranging. Proceedings of a Symposium held in Austin, Tex., U.S.A., 8–10 June, 1976.* 1977, XVII + 302 pp.
63. Giovanni G. Fazio (ed.), *Infrared and Submillimeter Astronomy. Proceedings of a Symposium held in Philadelphia, Penn., U.S.A., 8-10 June, 1976.* 1977, X+226 pp.
64. C. Jaschek and G. A. Wilkins (eds.), *Compilation, Critical Evaluation and Distribution of Stellar Data. Proceedings of the International Astronomical Union Colloquium No. 35, held at Strasbourg, France, 19-21 August, 1976.* 1977, XIV+316 pp.
65. M. Friedjung (ed.), *Novae and Related Stars. Proceedings of an International Conference held by the Institut d'Astrophysique, Paris, France, 7-9 September, 1976.* 1977, XIV+228 pp.
66. David N. Schramm (ed.), *Supernovae. Proceedings of a Special IAU Session on Supernovae held in Grenoble, France, 1 September, 1976.* 1977, X+192 pp.
67. Jean Audouze (ed.), *CNO Isotopes in Astrophysics. Proceedings of a Special IAU Session held in Grenoble, France, 30 August, 1976.* 1977, XIII+195 pp.
68. Z. Kopal, *Dynamics of Close Binary Systems*, forthcoming.
69. A. Bruzek and C. J. Durrant (eds.), *Illustrated Glossary for Solar and Solar-Terrestrial Physics.* 1977, approx. 216 pp.
70. H. van Woerden (ed.), *Topics in Interstellar Matter.* 1977, VIII + 295 pp.
71. M. A. Shea, D. F. Smart, and T. S. Wu (eds.), *Study of Travelling Interplanetary Phenomena.* 1977, XII+439 pp.
72. V. Szebehely (ed.), *Dynamics of Planets and Satellites and Theories of Their Motion. Proceedings of IAU Colloquium No. 41, held in Cambridge, England, 17-19 August 1976.* 1978, xii + 375 pp.
73. James R. Wertz (ed.), *Spacecraft Attitude Determination and Control.* 1978, xvi + 858 pp.